Combined Heat and Power

The Power Generation Series

Paul Breeze—Coal-Fired Generation, ISBN 13: 9780128040065
Paul Breeze—Gas-Turbine Fired Generation, ISBN 13: 9780128040058
Paul Breeze—Solar Power Generation, ISBN 13: 9780128040041
Paul Breeze—Wind Power Generation, ISBN 13: 9780128040386
Paul Breeze—Fuel Cells, ISBN 13: 9780081010396
Paul Breeze—Energy from Waste, ISBN 13: 9780081010426
Paul Breeze—Nuclear Power, ISBN 13: 9780081010433
Paul Breeze—Electricity Generation and the Environment, ISBN 13: 9780081010440

Combined Heat and Power

Paul Breeze

ACADEMIC PRESS

An imprint of Elsevier

Academic Press is an imprint of Elsevier
125 London Wall, London EC2Y 5AS, United Kingdom
525 B Street, Suite 1800, San Diego, CA 92101-4495, United States
50 Hampshire Street, 5th Floor, Cambridge, MA 02139, United States
The Boulevard, Langford Lane, Kidlington, Oxford OX5 1GB, United Kingdom

Notices
Knowledge and best practice in this field are constantly changing. As new research and experience broaden our
understanding, changes in research methods, professional practices, or medical treatment may become
necessary.

Practitioners and researchers must always rely on their own experience and knowledge in evaluating and using
any information, methods, compounds, or experiments described herein. In using such information or methods
they should be mindful of their own safety and the safety of others, including parties for whom they have a
professional responsibility.

To the fullest extent of the law, neither the Publisher nor the authors, contributors, or editors, assume any
liability for any injury and/or damage to persons or property as a matter of products liability, negligence or
otherwise, or from any use or operation of any methods, products, instructions, or ideas contained in the
material herein.

British Library Cataloguing-in-Publication Data
A catalogue record for this book is available from the British Library

Library of Congress Cataloging-in-Publication Data
A catalog record for this book is available from the Library of Congress

ISBN: 978-0-12-812908-1

For Information on all Academic Press publications
visit our website at https://www.elsevier.com/books-and-journals

  Working together
to grow libraries in
developing countries

www.elsevier.com • www.bookaid.org

Publisher: Joe Hayton
Acquisition Editor: Lisa Reading
Editorial Project Manager: Mariana Kuhl
Production Project Manager: Sruthi Satheesh
Cover Designer: MPS

Typeset by MPS Limited, Chennai, India

CONTENTS

CHAPTER 1

An Introduction to Combined Heat and Power

Much of the electricity that is generated on our planet is produced using heat engines that convert heat into electrical power. The heat for these heat engines comes from a variety of sources. Most is produced through the combustion of fossil fuels in steam plants, gas turbines plants, and reciprocating engines. Energy in nuclear power plants is also released in the form of heat which is used to drive a heat engine, usually a steam turbine, while biomass is often burned, too, to release heat.

The production of electricity in these plants from coal, oil, gas, and biomass is an inefficient process. While some modern combustion plants can achieve 60% energy conversion efficiency, most operate closer to 30% and smaller or older units may reach only 20%. The USA, which has a typical developed-world mix of fossil fuel based combustion plants, achieves an average fuel-to-end-user power plant efficiency of 33%, a level that has barely shifted for the past 30 years. Other countries would probably struggle to reach even this level of efficiency. Nuclear power plants are relatively inefficient too, with typical efficiencies of around 33%.

Putting this another way, between 40% and more than 80% of all the energy released in thermal power plants is wasted. The wasted energy emerges as heat which is dumped in one way or another. Sometimes, it ends up in cooling water which has passed through a power plant and then returned to a river or the sea, but most often it is dissipated into the atmosphere through a heat-exchanger. This heat can be considered a form of pollution.

Efficiency improvements can clearly curtail a part of this loss. A plant that has an efficiency of 60% will dump 40% less heat than a power station that achieves only 30% fuel-to-electrical efficiency. But even with the most efficient energy conversion system, a significant loss of energy is inevitable. Neither thermodynamic nor electrochemical energy conversion processes can operate even theoretically at anywhere near 100% efficiency and practical conversion efficiencies are always below the theoretical limit. So while technological advances

Combined Heat and Power. DOI: https://doi.org/10.1016/B978-0-12-812908-1.00001-8

may improve conversion efficiencies, a considerable amount of energy will always be wasted.

While this energy cannot be utilized to generate electricity, it can still be employed. Low grade heat can be used to produce hot water or for space heating,[1] while higher grade heat will generate steam which can be exploited by some industrial processes. In this way, the waste heat from power generation can replace heat or steam produced from a high value energy source such as gas, oil, or even electricity. This represents a significant improvement in overall energy efficiency.

Systems which utilize waste heat in this way are called combined heat and power or CHP systems. The term cogeneration is often used too while district heating and cooling (DHC) uses essentially the same technologies. Such systems can operate with an energy efficiency of up to 90% when heat usage is taken into account. This represents a major saving in fuel cost and in overall environmental degradation. Yet while the benefits are widely recognized, the implementation of CHP remains low.

Part of the problem lies in the historical and widespread preference for large central power stations to generate electricity. Large plants are efficient, and they are normally built close to the main transmission system so that power can be delivered into the network easily. They may also be sited close to a source of fuel. This will often mean that they are far from consumers that can make use of their waste heat.

If central power plants are built in or near cities and towns then they can supply heat as well as power by using their waste heat in district heating systems. Municipal utilities in some European and US cities have in the past built power plants within the cities they serve in order to exploit this market for heat as well as power but it is not an approach that has been widely adopted and environmental considerations makes building large power plants in cities more difficult today. There are also many examples of power plants being built close to industrial centers such that they can provide high grade steam for industrial use. In the main, however, large fossil fuel power plants simply waste a large part of the energy they release from the fuel.

District heating is a CHP application that is particularly important because there is a market for office and domestic heating in cities

---

[1]Heat can also be used to drive chillers and cooling systems. These are not considered separately here.

Table 1.1 District Heating Penetration by Nation	
Country	Penetration (%)
Iceland	95
Denmark	60
Estonia	52
Poland	52
Sweden	50
Czech Republic	49
Finland	49
Slovakia	40
Germany	22
Hungary	16
Austria	12.5
Netherlands	3
United Kingdom	2
Source: Wikipedia[2]	

across the globe. In spite of this, uptake is patchy. Table 1.1 shows penetration figures for the beginning of the 21st century in some European countries. As the figures show, levels range from 95% in Iceland, where geothermal heating is widely used, to 2% in the United Kingdom where there is no tradition of using district heating in towns or cities. Elsewhere, there is limited use of district heating in the USA and in Japan, and it is used sparingly in parts of China. Greater uptake, as more and more people move to cities, could lead to significant energy savings across the globe.

At a smaller scale, the situation is slightly better. At the distributed generation level, in particular, where power is generated either for private use of to feed into the distribution level of a power supply network, it is much easier to find local sources of heat demand that can be met at the same time as power is generated. This means that there are greater opportunities to achieve higher energy efficiency. Again, however, adoption of CHP technology is patchy.

In an energy-constrained and environmentally stressed world, energy efficiency represents one of the best ways of cutting energy use and reducing atmospheric emissions. The German government has

[2]https://en.wikipedia.org/wiki/District_heating.

estimated that 50% of its electricity could be supplied through CHP systems. There are economic advantages too that make greater use of CHP an extremely attractive proposition. In spite of these arguments, growth in the use of CHP has been painfully slow and it remains a major challenge for the electricity industry to achieve higher energy efficiency through the use of CHP.

The International Energy Agency has estimated that the adoption of CHP when new power generating capacity is built could reduce carbon dioxide emissions significantly. In a report published in 2008 it suggested that CHP could reduce emissions from new generation by around 4% in 7 years and roughly 10% in 22 years, as well as reducing the investment needed in the power sector by 7% over 25 years. In the second decade of the 21st century the proportion of total electricity generation associated with CHP plants is around 9%, a level that has remained static for several years.

THE HISTORY OF COMBINED HEAT AND POWER

The concept of CHP generation is not new. Indeed, the potential for combining the generation of electricity with the generation of heat was recognized early in the development of the electricity generating industry. In the USA, for example, at the end of the 19th century city authorities used heat from the plants they had built to provide electricity for urban lighting to supply hot water and space heating for homes and offices too. Steam can still be seen rising from manhole covers in some US cities, although the use of the technology has declined in recent years. These district heating schemes, as they became known, were soon being replicated in other parts of the world.

In the United Kingdom, around that time, a small number of engineers saw in this a vision of the future. Unfortunately their vision was not shared and uptake was slow. It was not until 1911 in the United Kingdom that a district heating scheme of any significance, in Manchester city center, was developed.[3] Fuel shortages after World War I, followed by the Great Depression made district heating more

[3]Combined Heat and Power in Britain, Stewart Russell in The Combined Generation of Heat and Power in Great Britain and the Netherlands: Histories of Success and Failure R1994: 29 (Stockholm: NUTEK, 1994).

attractive as electricity generation expanded in Europe during the 1920s and 1930s. Even here, however, take up was patchy. Nevertheless by the early 1950s district heating systems had become established in some cities in the USA, in European countries such as Germany and Russia, and in Scandinavia. In other countries like the United Kingdom, there was never any great enthusiasm for CHP and it gained few converts though a number of schemes were built after the Second World War as regions devastated by bombing were rebuilt.

This pattern of patchy exploitation has continued and the situation is complicated by the fact that it is almost impossible, economically, to build district heating infrastructure in modern cities that lack it. The centralization of the electricity supply industry must take some blame for this lack of implementation. Where a municipality owns its own power generating facility it can easily make a case on economic grounds for developing a district heating system. But when power generation is controlled by a centralized, often national body, the harnessing of small power plants to district heating networks can be seen as hampering the development of an efficient national electricity system based on large, central power stations—unless, that is, the CHP approach is already a part of the philosophy of the national utility.

Power industry structure is not the only factor. Culture and climate are also significant. So, while countries such as the United Kingdom failed to make significant investment in district heating, Finland invested heavily. Over 90% of the buildings in its major cities are linked to district heating systems and over 25% of the country's electricity is generated in district heating plants. Many Russian cities, too, have district heating systems with heat generated from large local power stations. Even some nuclear power plants in Russia are harnessed in this way.

District heating was—and remains—a natural adjunct of municipal power plant development. But by the early 1950s, the idea was gaining ground that a manufacturing plant, like a city, might take advantage of CHP too. If a factory uses large quantities of both electricity and heat, then installing its own power station allows it to control the cost of electricity and to use the waste heat produced, to considerable economic benefit. Paper mills and chemical factories are typical instances where the economics of such schemes are favorable and many such plants operate their own CHP plants.[4]

While this idea slowly gained ground, technological advances during the 1980s and 1990s made it possible for smaller factories, offices and even housing developments to install CHP systems based around small piston engine or gas turbine generating systems. In many cases, this was aided by the deregulation of the power supply industry and the introduction of legislation that allowed small producers to sell surplus power to the local grid. Since the middles of the 1990s the concept of distributed generation has become popular and this has also encouraged CHP.

Legislation in some countries has also helped encourage industrial and commercial CHP. In the USA, CHP district heating schemes were on the wane by the 1970s but introduction of the Public Utility Regulatory Policies Act in 1978 and the Energy Policy Act of 1992 both helped promote the use of CHP by mandating the sale of power by consumers to utilities. Fig. 1.1 shows how this led to an increase in the use of CHP from 1980 onwards so that by 2013, the electricity

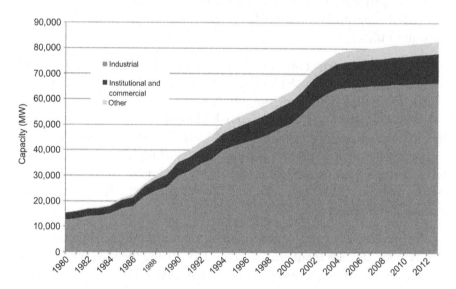

Figure 1.1 The growth of CHP in the USA between 1980 and 2013. Source: American Council for an Energy Efficient Economy.

---

[4]Applications of this type are frequently designated cogeneration rather than combined heat and power. The underlying premise is identical, however.

generating capacity in the USA associated with CHP had increased to just over 80,000 MW, from around 15,000 MW in 1980.

Recent concern for the environment now plays its part too. Pushing energy efficiency from 30% to 70% or 80% more than halves the atmospheric emissions from a power station on a per kWh basis. Thus CHP is seen as a key emission control strategy for the 21st century. But while environmentalists call for expanded use, actual growth has remained slow.

GLOBAL COMBINED HEAT AND POWER CAPACITY

CHP comes in many forms. These range from large industrial complexes with dedicated CHP plants, through urban district heating systems based on local power stations to small commercial and domestic systems that provide electricity, hot water and heating locally. This disparate array of CHP sources means that data on CHP uptake is rarely collated at a national level, making an estimate of total CHP usage extremely difficult to assess.

According to figures from the Worldwatch Institute just over 8% of the world's electricity generation includes cogeneration of one form or another[5]; the IEA puts the figure at around 9%. Most of this capacity, 80% according to the Worldwatch Institute, is associated with energy intensive industries such as paper, chemicals, oil refining and food processing. Most of the remainder is likely to be used to produce heating and hot water. The global electricity generating capacity in 2015 was 6400 GW according to the IEA. Using the IEA figure for CHP penetration of 9%, this would suggest that the global CHP capacity in 2015 was 576 GW.

A breakdown of the energy sources that contribute to global CHP is shown in Table 1.2. According to the figures in the table, natural gas was the most important fuel burnt in CHP plants, accounting for 53% of the total capacity across the world. Coal is the second most important source with 36% followed by oil with 5%. Renewable sources of energy that can be utilized in CHP are limited. There are some geothermal power plants that generate electricity and provide

[5]One-twelfth of Global Electricity Comes from Combined Heat and Power Systems http://www.worldwatch.org/node/5924.

Table 1.2 Breakdown of Global CHP by Fuel	
Fuel	Global CHP Proportion
Natural gas	53
Coal	36
Oil	5
Renewable sources	6
Source: Worldwatch Institute[6]	

Table 1.3 Combined Heat and Power Capacity in the European Union, 2005–15		
Year	CHP Electrical Capacity (GW)	CHP Heat Capacity (GW)
2005	101	n/a
2006	134[a]	n/a
2007	98	n/a
2008	100	n/a
2009	101	295
2010	83[b]	266
2011	106	266
2012	111	285
2013	113	279
2014	119	286
2015	120	304
[a]This figure includes a spuriously high German capacity.		
[b]Does not include capacity in Germany.		
Source: Eurostat[7]		

heat for district heating systems too. Biomass power plants are important in industries such as wood and paper where waste is burnt to generate power. Municipal solid waste and landfill gas are also used to provide CHP, although the absolute capacities are limited. Nuclear CHP is not included in Table 1.2 but there is a small nuclear CHP capacity in Russia and some Russian plants providing nuclear heating in other countries that were part of the former Soviet Union.

The adoption of CHP and district heating in different nations across the globe is uneven. Again according to the Worldwatch

[6]One-twelfth of Global Electricity Comes from Combined Heat and Power Systems http://www.worldwatch.org/node/5924.
[7]Eurostat Combined Heat and Power (CHP) data July 2017.

Institute, the regions that uses CHP most heavily are Western and Eastern Europe. In Western Europe, half of the CHP capacity is based on district heating systems. According to Cogen Europe, there are 5000 district heating systems in Europe which, together, account for 10% of the heating market.[8] The European Union (EU) is one region that has attempted to quantify CHP. Table 1.3 shows figures for the CHP capacity in the EU between 2005 and 2015. During this period both the electrical capacity and the heating capacity associated with CHP plants has risen. In 2005 there were 101 GW_e of electrical generating capacity in CHP plants (heat output in not recorded in the source for Table 1.3). This had risen to 106 GW_e in 2011 and 120 GW_e in 2015. Meanwhile heat capacity from CHP plants rose from 295 WW_{th} in 2009 to 304 GW_{th} in 2015. Based on the global estimate for CHP electrical capacity of 325 GW, above, the EU capacity accounts for around 37% of the global total.

A more detailed breakdown of CHP electricity and heat production for the 28 nations of the EU, plus Norway, are shown in Table 1.4.

Table 1.4 CHP Production in the European Union, 2015		
Country	CHP Electricity Production (GW h)	CHP Heat Production (PJ)
Belgium	12.5	104.4
Bulgaria	2.9	31.9
Czech Republic	11.8	106.0
Denmark	11.6	93.3
Germany	78.8	699.9
Estonia	1.2	12.5
Ireland	2.1	12.6
Greece	2.0	10.9
Spain	22.7	120.3
France	13.9	155.0
Croatia	0.8	10.0
Italy	39.5	213.2
Cyprus	0	0.2
Latvia	0	12.4
Lithuania	2.5	12.4
Luxembourg	1.5	2.4

(Continued)

[8]http://www.cogeneurope.eu/district-heating_270.html.

Table 1.4 (Continued)		
Country	CHP Electricity Production (GW h)	CHP Heat Production (PJ)
Hungary	0.4	24.4
Malta	4.1	–
Netherlands	–	189.6
Austria	9.0	105.9
Poland	26.5	238.6
Portugal	6.5	59.3
Romania	5.6	51.0
Slovenia	1.2	10.4
Slovakia	21.1	27.3
Finland	21.7	242.4
Sweden	13.7	151.4
United Kingdom	19.4	124.2
Norway	0.4	9.7
Source: Eurostat[9]		

These figures from Eurostat show that Germany is the largest user of CHP in the EU with 670 PJ of thermal production in 2015. CHP electrical generating capacity in Germany (not shown in the table) was 37 GW and electrical production from these plants was 78.8 GW h. Finland, Poland and Italy are also large heat producers while the Netherlands has the second largest electrical generating capacity associated with CHP at 9 GW. Poland also has 9 GW of CHP electrical generating capacity and produced 26.5 GW h of power from these plants. At the other end of the scale, Cyprus has no CHP and there are very limited capacities in Croatia, Slovenia, Greece, Ireland, and Bulgaria. Further figures from Eurostat show that in 2015 the fuel breakdown for CHP in the EU comprised 44% natural gas, 21% renewables, 18% solid fuels and peat, 13% other fuels, and 5% oil and oil products.[10]

The USA has a relatively large installed base of CHP plants (see Fig. 1.2). According to the US Department of Energy, there was around 82.7 GW of CHP capacity at the end of 2014. Most of this was located at more than 4400 industrial and commercial facilities across the country and the aggregate accounted for 8% of US

[9]Eurostat Combined Heat and Power (CHP) data July 2017.
[10]Rounding error makes the sum of these sources 101%.

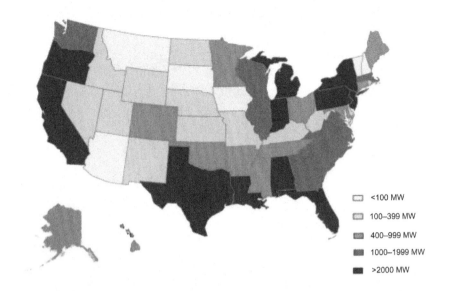

Figure 1.2 CHP capacity in the USA, by state. Source: US Department of Energy.

generating capacity and 12% of US electricity production. Many of the US plants are relatively large is size, with more than 85% of plants over 50 MW in electrical generating capacity.[11]

Estimates of the CHP capacity in China vary widely. Worldwatch institute has put it at around 28 GW while a presentation from the US DOE has suggested more than double that figure. Most is coal-fired and based on small power plants which are likely to be polluting an inefficient. The Chinese government has been targeting such plants for closure so the actual capacity could be lower than the US DOE estimate.[12] Against this, a recent directive is aimed at the expansion of natural gas-fired cogeneration. Russia has historically maintained a large CHP capacity, most of it used for district heating. However much of this infrastructure is aging, dating back to the early days of the Soviet era, so while there may be as much as 70 GW of capacity, the amount that is fully operational is more difficult to gauge.

[11]One-twelfth of Global Electricity Comes from Combined Heat and Power Systems http://www. worldwatch.org/node/5924.
[12]Combined Heat and Power in China: Lessons on Technology Adoption, M. Evans, Pacific Northwest National Laboratory, GTSP Technical Review Meeting, May 28, 2008.

CHAPTER 2

The Combined Heat and Power Resource: Estimating the Potential

Combined heat and power (CHP) is a technology (or a range of technologies) not an energy source. Even so it can be quantified as if it were a resource available to be exploited. Whenever fuel is consumed to generate electrical power using a heat engine, waste heat will also be produced. By looking at the amount of electricity that is produced using this type of generation system, globally, it is possible to put a maximum figure on the amount of heat energy that is potentially available from such power plants. Figures from the International Energy Agency (IEA) show that in 2014, the world produced 23,816 TW h of electrical power, 77.3% of which came from power plants that either burnt fossil fuels or were reliant on heat from a nuclear reactor to produce electricity.[1] Assuming that these have a global average efficiency of 33%, 36,820 TW h of heat energy was potentially released into the environment from those power plants. This represents the upper limit on the amount of energy that might have been available during that year for CHP applications.

Of course, there are technical limits on the amount of this energy that can be exploited, and some of it is already used. Furthermore, as renewable sources of electricity provide more and more electricity, the amount produced using heat engines will fall and so will the amount waste heat available. Even so, it is likely that there will be a massive amount of heat energy that could potentially be exploited, even by the end of the 21st century.

The exploitation of this energy using CHP technology offers significant benefits in all parts of the globe. These benefits are both economic and environmental. Economically, the use of CHP leads to more efficient use of fuel resources which in turn provides cheaper energy.

---

[1]Key World Energy Statistics, IEA, 2106.

Combined Heat and Power. DOI: https://doi.org/10.1016/B978-0-12-812908-1.00002-X

Analysis by the IEA suggests that growth in CHP markets can lead to a small reduction in the end user cost of electricity, if the savings are allocated in this way.[2] (As the IEA has pointed out, many other analyses suggest that the use of CHP raises the cost of electricity to end users.) Meanwhile, the better use of the heat released during power generation can provide significant economic benefits to recipients of that heat energy.

A further potential economic benefit is a reduction in the investment required in the power sector through the more efficient use of existing heat and power generating capacity. Since CHP offers local heat and power, the amount of centrally generated power required is reduced, leading to lower investment in transmission and distribution system expansion, while the use of local CHP plants lowers the overall investment needed in central power stations. The IEA estimated that promoting CHP could reduce the amount of power sector investment required between 2005 and 2030 by 7%.

Perhaps more importantly, CHP offers major environmental advantages, cutting greenhouse gas emissions and helping fight global warming. In the European Union, for example, it has been estimated that CHP delivered 15% of greenhouse gas emission reductions achieved between 1990 and 2005. If these gains were replicated across the globe, significant progress would be achieved in the fight against carbon dioxide emissions.

Unfortunately, estimates also suggest that the amount of CHP, as a proportion of total global electricity generation, is not advancing. As noted in Chapter 1, estimating the amount of CHP in use across the globe is difficult, but the most far reaching estimates carried out by the IEA show that the level had stabilized at around 9% of total generating capacity towards the end of the first decade of the 21st century.

In order for this penetration level to increase, ways need to be found of exploiting the remaining potential. The extent to which this is possible depends on the demand for heating and cooling in each country and region (a waste heat source can be used for cooling as well as heating). While this will vary, there is clearly more that can be done. For example, five countries in the world have been able to expand

[2]Combined Heat and Power: Evaluating the benefits of greater global investment, International Energy Agency, 2008.

their CHP use to account for between 30% and 50% of total power generating capacity. As the average quoted above suggests, most nations exploit less than 10% of their generating capacity in this way.

GLOBAL CHP POTENTIAL

The most reliable recent estimate for the amount of CHP in use across the globe is based on a survey by the IEA carried out during the first decade of the 21st century.[3] This suggested that the percentage of global electricity generation associated with CHP was around 9%. The IEA concluded at the time that CHP penetration had stagnated at around this figure. Meanwhile, its survey found that 37 nations examined in detail, including most of the major world economies, had an aggregate electrical generating capacity from CHP plants of 333 GW (see Table 2.1). In 2006, the global electrical generating capacity, according to the World Energy Outlook 2008, was 4344 GW implying that there would have been a total of around 390 GW of CHP in use in the middle of the decade when this study was carried out. On this basis, the nations in the survey accounted for around 85% of the global total. Generating capacity in 2015 was 6400 GW, and if penetration has remained the same, then this global capacity would indicate 576 GW of electrical generating capacity was linked to CHP plants.

The European Union as a region has traditionally been the largest user of CHP. It had around 120,000 MW of electrical generating capacity associated with CHP[4] in 2015. Within the region, the largest national fleet is found in Germany which had 37 GW of electrical capacity associated with CHP in 2015. This is significantly higher than the 21 GW shown in Table 2.1 from the first decade of the century, consistent with the continuing growth in CHP capacity implied above. On the other hand, the USA is shown with nearly 85,000 MW of CHP capacity in Table 2.1. This has barely changed in a decade. Other important CHP users include Russia with 65,000 MW and China with 28,000 MW. (The capacity in Russia is probably unchanged since the first decade of the century but that in China may well have increased significantly.) India with 10,000 MW, Japan with 9000 MW, Taiwan

---

[3]While these figures are now old, there has been little change in the ensuing decade in some countries, including the USA. Penetration has increased significantly in others, such as Germany and China.
[4]Eurostat Combined Heat and Power (CHP) data July 2017.

Table 2.1 Estimated CHP Capacity for 37 Countries			
Country	CHP Capacity (MW)	Country	CHP Capacity (MW)
Australia	1864	Japan	8723
Austria	3250	Korea	4522
Belgium	1890	Latvia	590
Brazil	1316	Lithuania	1040
Bulgaria	1190	Mexico	2838
Canada	6765	Netherlands	7160
China	28,153	Poland	8310
Czech Republic	5200	Portugal	1080
Denmark	6600	Romania	5250
Estonia	1600	Russia	65,100
Finland	5830	Singapore	1602
France	6600	Slovakia	5410
Germany	20,840	Spain	6045
Greece	240	Sweden	3490
Hungary	2050	Taiwan	7378
India	10,012	Turkey	790
Indonesia	1203	United Kingdom	5400
Ireland	110	USA	84,707
Italy	5890		

Some of the figures in this table contradict those in Table 1.4. The European Union figures in that table are from 2015: figures in this table are from the first decade of the 21st century.
Source: IEA[5]

with 7000 MW, and the Netherlands with 7000 MW also make good use of CHP.

The absolute CHP capacities presented in Table 2.1 do not tell the whole story. When figures for the share of CHP in total generation are examined, the story is often one of lost opportunities. Of the main global economies, only Russia has more than 30% (the actual figure is 31%) of its generating capacity exploited for CHP. Germany and China exploit around 12% each; all the remaining large economies use less than 10%. There are nations that do better. Denmark has around 50% (some reports claim 60%) of its generating capacity associated with CHP, Finland has just under 40%, and Latvia and the Netherlands around 30%. However, these are the

[5]Combined Heat and Power: Evaluating the benefits of greater global investment, International Energy Agency, 2008.

Table 2.2 Global Final Energy Consumption Breakdown		
Energy Use	OECD (%)	World (%)
Heat	37	47
Electricity	21	17
Transport	32	27
Nonenergy use	10	9
Source: IEA[6]		

exceptions. Nevertheless, these figures show that a much larger proportion of generating capacity can be exploited for CHP that has been achieved across most of the globe.

One way of assessing the unexploited potential is to assume, arbitrarily, that the figure from Denmark of 50% penetration represents an upper limit on the amount of electrical generating capacity that can be exploited for CHP. If this is applied to the IEA global generating capacity figure of 6400 GW in 2015, it implies that globally there could be 3200 GW of CHP capacity. This top-down approach is probably extremely optimistic and will likely overestimate the actual potential. The alternative is to start from the bottom and estimate the prospective heat demand in each country of region and then convert this into potential CHP capacity. This is not easy, either.

One starting point is to look at what proportion of global energy resources are used simply to produce heat each year. The results are startling. Heat production is the largest consumer of energy resources across the globe. Table 2.2 shows figures from the IEA breaking down global final energy consumption by type. The figures show that across the globe 47% of the energy resources consumed are used simply to provide heat. In comparison, only 17% is used to generate electricity while 27% is used for transport. Within the developed nations of the Organization for Economic Cooperation and Development (OECD) heat accounts for 37% of the final energy consumption somewhat lower but still the largest single category.

The use of energy sources to produce heat directly can be further subdivided. Again, based on IEA figures, industry is the largest single consumer, accounting for 43% of the total globally and 44% in OECD

[6]Cogeneration and Renewables: Solutions for a low-carbon energy future, IEA 2011.

Table 2.3 Global CHP Potential for Selected Regions and Countries	
Country/Region	Estimated CHP Potential (GW)
China	332
USA	150
EU	122
Russia	87
India	85
Japan	26
Canada	25
Brazil	15
South Africa	12
Mexico	7
Source: GE[7]	

nations. Meanwhile the buildings sector, which aggregates the residential, commercial, and public services sectors, accounts for 50% of final energy use globally and 52% within the OECD. A large proportion of the residential component of this heat use is likely to be the use of traditional biomass for cooking. Outside the OECD, this practice remains widespread. Even so, in principle, a large part of the buildings sector heat could be supplied from CHP systems that produced electricity as well.

Clearly, then, there is a large demand for heat energy across the globe. Converting this demand into quantitative figures for national CHP potentials is not an easy task but efforts have been made, usually at a national level as part of a climate change strategy. Table 2.3 presents one attempt to quantify the potential for a limited number of nations where figures are available. As they show, it has been estimated that the CHP potential in China is 332 GW, and in the USA, the figure is 150 GW. For the EU, the estimate is 122 GW, a level that has already been reached. In Russia there is estimated to be potential for 87 GW of CHP while in India the figure is 85 GW. The other countries listed include Japan with potential for 26 GW of CHP,

[7]Combined Heat & Power (CHP) US Market Overview, Daniel A Loero, GE. The figures are drawn from the International Energy Agency and from the US Energy Information Administration.

Canada where the potential is 25 GW, Brazil with 15 GW, South Africa with 12 GW and Mexico with 7 GW.

There are other figures and surveys which present a different picture. For example, according the European Commission Energy Roadmap 2050, the share of electrical production from CHP was expected to rise from 473 TW h in 2005 to around 1050 TW h in 2030, before declining to 700 TW h in 2050 as renewable sources have a larger impact. Meanwhile the IEA survey suggests that total CHP potential in the EU ranges from 150 to 250 GW. In Japan, the potential in 2030 could be 29 GW.

Taking a broader approach, the IEA has also estimated that for a group of nations it labels the G8 + 5,[8] the share of CHP in electricity production could rise to 24% by 2030, from an estimated 11% at the end of the first decade of the 21st century. For individual nations, much depends on the local circumstances. For example Brazil generates much of its electrical power using hydropower plants which have no potential for CHP.

While these and other analyses suggest that much more use could be made of our energy resources if CHP was used intelligently and more extensively, this is not a new conclusion. For several decades now energy agencies and other organizations have been trying to promote the use of CHP as a way of improving energy efficiency and energy economy. In most countries, the results of these efforts remain limited.

[8]The G8 group of nations—Canada, France, Germany, Italy, Japan, the United Kingdom, USA, Russia—plus five leading emerging economies, Brazil, China, India, Mexico, and South Africa.

CHAPTER *3*

Combined Heat and Power Principles and Technologies

Combined heat and power (CHP) can be defined simply as the simultaneous use of electricity and heat from a single energy source. This energy source may be a fossil fuel such as coal or natural gas, it can be nuclear fuel or it might be a renewable fuel; geothermal energy, biomass and solar thermal power can all potentially form the basis for a CHP system.

Virtually, all CHP systems are based around the use of heat engines to generate electrical power. The main exception is the fuel cell which generates power electrochemically. Heat engines that are used for electricity generation include steam turbines, gas turbines and a variety of reciprocating engines. All these engines can be understood in terms of thermodynamic principles established during the 19th century, principles which underpinned their development. Since the beginning of the 20th century their use has grown so that today, between them, they provide the largest part of the world's electricity.

Electricity and heat are both energy sources, but their properties are different. Electricity can be delivered to every household and organization in a nation because it is relatively easy to transport swiftly and efficiently. In contrast, heat energy cannot conveniently and economically be transported with anywhere near the same ease. Although it is possible to transport heat over long distances most heat is used locally, close to the point at which it is generated. This dictates one of the key elements of CHP system design. A fundamental principle for optimum CHP efficiency is to identify a demand for heat and then design a plant with a heat output to meet this demand. Electricity from the plant can, in this sense, be considered as a valuable by-product.

When looking at CHP technology, it is useful to identify two different types of CHP system. In the first, often called a topping cycle, the fuel supplied to the system is used primarily to generate electricity

Combined Heat and Power. DOI: https://doi.org/10.1016/B978-0-12-812908-1.00003-1

while the heat that is left after electrical power has been generated is used in an ancillary application. Depending upon the type of generating system, the heat energy from the electricity generating system might be high grade heat that is suitable for raising steam or for use directly in an industrial process, or it might be lower grade heat that is only suitable for space heating and hot water production.

The second type of CHP system is called a bottoming cycle. In this type of system the fuel is used first to produce heat. The heat will normally be exploited in an industrial process that requires very high temperatures or vast quantities of heat. Iron smelting is a typical example. Any energy not used in the process is then used to generate electricity. Bottoming cycle CHP is also called waste energy recovery or waste heat recovery. The energy recovered for electricity generation may be in the form of heat but it may also be combustible gases such as those produced during iron production. Some electricity generating systems can also exploit high pressure exhaust gases from a process, using the pressure drop to drive an engine. Bottoming cycle CHP systems are particularly attractive because they can provide electricity without the need consume additional fuel.

COMBINED HEAT AND POWER APPLICATIONS

While, ideally, a CHP system is designed around the heat demand in a particular location in many cases this is not how they evolve. Many designs are compromises that balance economic considerations with the optimum energy balance. The availability of off-the-shelf components may also play a part in determining the final outline of a system. In other cases the principle need may be for electricity and the use of heat is a secondary consideration. Even so, it is always the heat demand that will determine the location and outline of any CHP system. For without heat demand, there can be no CHP.

There is a further, practical consideration when addressing CHP development. The heat demand around which the CHP plant is constructed must be durable. So while CHP installations can range from single home heat-and-electricity units to municipal power stations supplying heat and power to a city, from paper mills burning their waste to provide steam and heat to large chemical plants installing gas turbine-based CHP facilities, they all share one theme. Ideally the heat and electricity from a CHP plant will be supplied to the same users. While this is not an absolute requirement it is a pragmatic principle for a successful CHP scheme.

If a heat and power plant is supplying both types of energy to the same users, be they an industrial plant or households, then the economics of the plant will remain sound so long as the customers remain. However, if a plant supplies electricity to one customer and heat to another, its economic viability can be undermined by the loss of either. Part of this risk is mitigated if a plant can export electricity directly into the grid while selling its heat locally but in general the economics of CHP will be most soundly based where the same customers take both.

Examples of how this can be achieved exist at all levels within the electricity market from the smallest electricity consumer to the largest. At the very bottom of the scale, many home owners buy electricity from a utility and either use this directly as a heat source or purchase natural gas to provide hot water and space heating. However, it is now possible to install domestic CHP systems based on microturbines, Stirling engines or fuel cells that will replace a domestic hot water system and at the same time generate electricity for use by the household, with excess power perhaps being sold to the local utility. While such systems remain costly, they are likely to become more cost effective in time.

On a larger scale, a microturbine or a reciprocating engine burning natural gas can be used to supply both electricity and heat to an office building, a large block of apartments or a small commercial or industrial enterprise. Such systems are widely available and can be installed in urban environments with ease. The system may be connected in such a way that excess power can be exported to the grid although the sizing of such systems is normally based on heat demand; power demand will often be higher than the system can supply alone but this can be topped up from the grid.

At the top end of the capacity scale, a municipal power plant based on a coal-fired boiler or a gas turbine can provide electricity for a city and heat for that city's district heating system. New systems of this type are difficult to install in existing cities but opportunities do arise where major redevelopment takes place and new urban housing schemes can be built with district heating too.

In all these cases, the need to supply electricity is usually the primary driving force but the use of heat from the power generation facility improves the economics considerably. Similar opportunities exist in industry but in many industrial cases the situation is reversed and it is heat or steam that is the primary requirement with electricity production a secondary consideration. However the same arguments apply. There are many industrial processes that require a source of heat and all industrial plants will use some electricity. Often, the two can be combined to good economic effect once the benefits are recognized. So where, in the past, a paper manufacturer would have installed a boiler to supply heat while buying power from the grid, now the same manufacturer is more likely to install a CHP plant, often fired with waste produced by the paper manufacturing process. Chemical plants often require a supply of high grade heat too, as do some refineries. Food processing plants may require large heat supplies as well.

Such instances represent the ideal but a good match of heat and power demand is not always possible at this level. However, provided the scale is large enough, such plants may be able to install a grid connected power plant that exports electricity while using heat generated from the same plant to supply its industrial needs. Alternatively, a large CHP plant can be built to supply an industrial site supporting a range of different industries, some with large heat requirements, others with demand for electrical power.

Of all these applications, domestic heat consumption remains the biggest challenge to the expansion of CHP. Where district heating networks exist, a good balance between domestic heat and electricity demand is possible. When there is no district heating network, the main options are either power stations meeting the electricity demand only, or domestic CHP systems. The latter are the only solution for many households but their installed base is still very small and cost is high. If costs can be brought down then such systems offer a real chance of a significant change in global energy usage.

CHP TECHNOLOGY

As already outlined, there are two types of CHP configuration to consider when analyzing CHP systems, a topping cycle and a bottoming cycle. A topping cycle takes its name from the fact that it the electricity

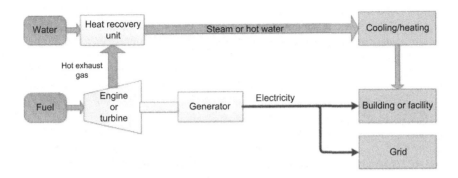

Figure 3.1 Schematic of a CHP topping cycle. Source: US Energy Information Administration.

generating system that is the primary, or top, user of an energy source. Other energy users in the system—in this case heat users—must make do with what is left over. A typical topping cycle is illustrated in Fig. 3.1. In this type of system, an electrical generating system such as a gas turbine, reciprocating engine (or fuel cell) is used to generate electricity and then energy that is rejected by the generating system as waste heat is captured and used to provide steam, process heat or for space heating and hot water. In general, with this type of system the heat that is available after power generation is relatively low grade. However, it is possible to design systems of this type which can provide high grade heat or steam by adjusting the overall balance between heat and power.

The second type of CHP configuration is the bottoming cycle. In this case, the electricity generating system is last in line, or at the bottom, and must make do with energy left after the primary heat user has finished. In this configuration a fuel is used first to provide heat for a furnace or industrial process. Heat energy that is vented or exhausted from this process is then used in a heat engine based power generation system to provide electricity. This is shown schematically in Fig. 3.2. In order for a bottoming cycle to be economically viable the heat energy that emerges from the industrial process must be of sufficient quality to be able to heat a thermodynamic fluid and drive a turbine. The latter will most often be a steam turbine. Where the heat energy is at a relatively low temperature it may be possible to use an organic Rankine cycle turbine which can convert lower grade heat into electricity. There are also instance where the energy to be recovered in a bottoming cycle system is in the form of combustible gas, such as

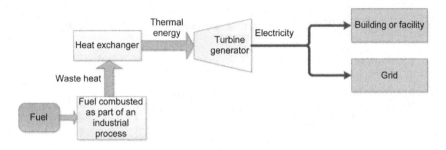

Figure 3.2 Schematic of a CHP bottoming cycle. Source: US Energy Information Administration.

from a blast furnace, or as a high-pressure gas stream which can be used to drive a gas turbine.

In addition, there are CHP systems that bridge the gap between the two. A typical example would be an industrial steam turbine CHP system which generates steam to drive a steam turbine, but with high pressure steam taken from the input or part way through the steam turbine to provide industrial process heat. This type of system is common in some industries.

All the types of power generation technology that are based on heat engines are capable of being integrated into a CHP system. Thus, all fossil fuel fired power plants and biomass power plants can be adapted to create CHP systems. In addition, electrochemical fuel cells can be an excellent source of heat as well as power. Among renewable technologies, solar thermal power plants can provide heat in addition to electricity if necessary and geothermal energy has been widely exploited for both electricity generation and district heating. Other renewable technologies such as wind, hydropower and marine power involve no heat generation. However, solar photovoltaic power generation can, in principle at least, be exploited in conjunction with solar heating because the solar cell only uses a part of the incident light and rejects most of the heat-bearing infra-red radiation. Nuclear power, too, can be exploited for heat and power generation although its use is rare outside Russia and countries of the former Soviet Union.

While this range of technologies offers a wide choice for a CHP plant, the type of heat required from a CHP application will often narrow the choice of technology. If high quality steam is demanded then a bottoming cycle may be the first choices. If a topping cycle is preferred, then a source of high temperature waste heat will be needed. This can be taken from a steam turbine-based power plant, it can be generated using the exhaust of a gas turbine and it can be found in a high temperature fuel cell. Other generating systems such as piston engines or low temperature fuel cells are usually only capable of generating hot water, and perhaps low quality steam.

For smaller applications and where only hot water is needed a reciprocating engine, microturbine or low temperature fuel cell might offer the best match. Again however, the mode of operation will determine the optimum choice. If the unit is to supply power to a particular consumer or group of consumers, with its output following their demand, then a generating unit that can operate efficiently at different load levels such as a piston engine or fuel cell will probably be the best solution. However if it is going to provide base-load generation then part load efficiency will be of less significance.

Finally, location will be important. It will not be possible to install some types of CHP plant in urban areas because of the emissions and the noise they generate. This will therefore limit the technologies available for use in this situation.

With a bottoming cycle CHP system, the amount of heat available is determined by the industrial process that requires it. For topping cycle CHP systems the quantity of heat available is determined, in part at least, by the generating technology and it will vary from technology to technology.

Table 3.1 gives typical energy conversion efficiency ranges for modern fossil fuel burning power plants. Modern high efficiency coal-fired power plants can operate at between 38% and 47% efficiency although

Table 3.1 Typical Power Plant Energy Conversion Efficiencies	
Type of Plant	Efficiency
Conventional coal-fired power plant	38% to 47% for modern high efficiency coal plants
Heavy industrial gas turbine	Up to 42%
Aeroderivative gas turbine	Up to 46%
Fuel cell	25% to 60% depending upon type
Natural gas-fired reciprocating engine	28% to 42%
Diesel engine	30% to 50%

there are many that are much less efficient. However high efficiency coal-fired plants produce little usable waste heat unless overall efficiency is compromised since the steam exiting the steam turbine is generally at a very low temperature and pressure. Gas turbines provide more flexibility while offering a similar electrical energy conversion efficiency since their exhaust gases can provide high grade heat.

Where an even greater level of flexibility is required, it is possible to design a plant to produce less electricity and more heat that the efficiency figures in Table 6.1 suggest. Some technologies are amenable to this strategy. Others are not. Most flexible are boiler/steam turbine plants but gas turbine CHP units can easily be adapted in this way too.

PISTON ENGINES

Two types of piston engine are regularly used for power generation, diesel engines and spark ignition gas engines. The former are the most efficient and can operate up to 50% fuel to electrical conversion efficiency. Spark ignition gas engines achieve around 42% efficiency at best. However the latter are cleaner than diesel engines which usually require extensive exhaust gas cleaning.

Piston engines can provide heat for hot water and space heating but rarely provide heat for processes that require higher grade heat although some larger engines can produce medium pressure steam. Heat is captured from engine cooling systems and from the exhaust gases. Piston engines suitable for CHP applications come in sizes ranging from a few kilowatts to several megawatts. They are often used to provide both electricity and heat for hot water and space heating in commercial and some large residential situations. The engines are good a load following with little fall in efficiency down to 50% load. This makes them more flexible than some other types of CHP prime mover.

STEAM TURBINES

Steam turbines are one of the workhorses of the power generation industry. Large steam turbines can achieve 47% efficiency. Smaller units that are likely to be used in CHP systems will have lower efficiency, though this need not be a problem. Steam turbines for large coal-fired or nuclear power plants will usually have multiple units to provide an output often in excess of 1000 MW. Single steam turbines of up to 250 MW are available while for small-scale applications, units as small as 50 kW can be found. With this wide capacity range it is normally possible to find a unit suitable for any CHP application.

A simple steam turbine CHP system needs a boiler and steam generator to raise the steam to drive the unit. The fuel can be any fossil fuel as well as biomass fuels. For highest electrical conversion efficiency, the steam will be condensed at the exit of the steam turbine, creating the largest pressure drop possible across the turbine blades. However, for CHP applications, steam is often taken from the steam turbine exhaust to use in a process. Such turbines are called back-pressure turbines because their exhaust pressure is not the minimum. In addition, it is possible to design a steam turbine CHP system in which steam is extracted at one or more points along the turbine. Turbines with this capability are called extraction steam turbines.

An alternative configuration is to take steam directly from the plant boiler to use in an industrial process and then use the remaining energy to generate electricity, much in the manner of a bottoming CHP system. The flexibility of steam turbines in CHP applications makes them ideal for many types of industrial process.

GAS TURBINES

Gas turbines are used in a range of power generation duties. Sizes range from under 1 MW to nearly 400 MW. Modern gas turbines have efficiencies of up to 46%. They are normally designed as stand-alone engines that burn a fuel directly to provide mechanical drive for a generator. Gas turbines usually burn natural gas today, but they can be fired with liquid fuels as well as a range of gases derived from biomass.

As with a steam turbine, the highest efficiency is extracted from a gas turbine when the pressure and temperature drop across its blades is a maximum, implying the lowest possible exhaust gas temperature and pressure. Even under these conditions the exhaust gas temperature will be high enough to raise medium or high pressure steam for CHP use. In a CHP system where more thermal energy is required, the efficiency of the gas turbine can be compromised by allowing the gases to exit at a higher temperature.

Many simple gas turbine CHP systems use a gas turbine to generate power and then exploit the hot exhaust gases to provide process heat, steam or heating. However some can be fitted with an additional steam turbine to generate more electrical power, but with reduced heat output. Alternatively, the heat output can be boosted by using some form of supplementary firing after the exhaust gases exit the turbine. Most gas turbine CHP systems are designed to provide industrial heat.

There is another type of gas turbine, called a microturbine, that is much smaller than standard industrial turbines. Sizes typically range from 3 to 250 kW. These units are designed specifically for commercial and domestic use and many are packaged as CHP systems, capable of providing both electricity and hot water.

FUEL CELLS

The fuel cell is an electrochemical device, similar in concept to a battery, that uses an externally supplied fuel to generate electrical power. There are a range of fuel cell types, some that work at low temperature and some that operate at high temperatures. Cells which operate at the lowest temperatures are generally not useful in CHP systems but most other types can be adapted for CHP use. One popular design, called the phosphoric acid fuel cell, has been marketed as a packaged CHP unit by some manufacturers. These cells operate at around 200°C and can provide heat sufficient for hot water and space heating.

Other fuel cells operate at temperatures between 600°C and 1000°C. These types of cell can potentially provide high grade heat that can be used to raise steam or for industrial processes. They can also be used for hot water and heating. Fuel cells are very clean and quiet and are ideal candidates for CHP in urban areas. Two types of fuel cell have been developed for use in domestic CHP systems.

NUCLEAR POWER

Nuclear power plants exploit nuclear fission reactions to generate heat that is then used to raise steam and drive a steam turbine. The plant configuration is virtually identical to that of a fossil fuel steam turbine plant, but with the boiler replaced by a nuclear reactor. In the same way as the fossil fuel technology can be used for CHP applications, so too can the nuclear technology.

Nuclear CHP plants have most normally been used for district heating applications although they can provide industrial heat too. The nuclear reactor normally generates steam that is at lower temperature and pressure than a standard fossil fuel power plant and so this is better suited to applications where the heat requirement is not extreme. The use of nuclear power for heat has been exploited in Russia and there are some plants in other eastern European nations that were once part of the Soviet Union. However this type of technology has not, so far, been widely adopted in the developed countries of the western world.

GEOTHERMAL POWER

Geothermal power generation relies on the heat from an underground reservoir of brine. The fluid is pumped to the surface and then used to generate power before, in the best designed systems, being reinjected into the underground reservoir. There is also a long history of brine from underground reservoirs being used for district heating.

There are many underground reservoirs of brine that are not hot enough for power generation; if these are exploited, it is normally for heating alone. Others, in which the brine is at a higher temperature can be used for both. The configuration will normally be that of a topping cycle with the hot brine being used first to drive a heat engine, before the remaining heat used to heat water for a district heating system. Geothermal energy is economical where there are good geothermal resources, but these are limited, geographically.

SOLAR THERMAL POWER

A solar thermal power plant uses the heat energy from the sun as an energy source to drive a heat engine. In order to make this effective, the solar energy must be collected over a wide area and concentrated so that a temperature sufficient for economic electricity generation can be achieved. There are a range of solar thermal technologies that use different types of concentrators to provide energy for power generation.

In principle the heat energy collected in this way can also be exploited in a CHP topping cycle, with the heat used first to provide electricity before the waste heat is used for heating. This is not a technology that has been exploited commercially. However, solar energy is commonly used directly to heat water using rooftop solar panels through which water is circulated.

CHAPTER 4

Piston Engine Combined Heat and Power Systems

The piston engine, or reciprocating engine, is probably the commonest form of heat engine to be found across the globe. Engines of this design provide the motive force for virtually all road vehicles in use today as well as powering trains and many ships. In addition, they are used for a variety of small devices such as lawn mowers and chain saws. The same engines are commonly employed as sources of electrical power too. Units for power generation can be as small as a few kilowatts or as large as 80 MW. The largest engines are also among the most efficient prime movers available for producing electrical power.

There are two broad categories of piston engine that can be used for power generation, the spark ignition engine and the compression ignition, or diesel engine. These differ in the way combustion of the fuel is initiated, and this leads to differences in both efficiency and emission levels from the engines. Spark ignition engines are generally cleaner but less efficient while diesel engines have higher efficiency but produce larger quantities of atmospheric emissions. One subgroup of spark ignition engines is the natural gas spark ignition engine. These engines have become popular for both power generation duties and for combined heat and power (CHP).

Both diesel and spark ignition engines can be adapted for CHP. Systems based on these engines are always topping cycles with the engine providing electrical power. Waste heat is then captured from the exhaust and cooling systems of the engines and used to provide heat energy for heating and hot water. Some larger engines can also be adapted for medium pressure steam production.

In addition to the two main types of piston engine, there is a third type that can be used for power generation, the Stirling engine. This engine is unique among commonly used reciprocating engines in that

Combined Heat and Power. DOI: https://doi.org/10.1016/B978-0-12-812908-1.00004-3

it is an external combustion engine in which the heat is applied to the outside of the engine rather than combustion of fuel taking place within the engine. Stirling engines have been used in solar thermal power plants and can be used in domestic CHP systems too.

The range of piston engines available today makes it relatively easy to find an engine to suit any particular CHP application. Small engines are mass-produced for cars while larger engines are designed for trucks and buses. The largest engines of all are marine and locomotive engines. All can be adapted for cogeneration. In most cases, engine CHP systems will be packaged so that all is required is to supply a fuel and connect the electrical and heating outputs to the local demand. Bespoke engine CHP systems are also available, at higher cost.

ENGINE TYPES

Piston engines suitable for CHP applications are primarily of two types, spark ignition engines and compression ignition engines. Within each category there are also two different cycles, the two-stroke cycle and the four-stroke cycle. The latter is the most common in engines used for power generation although some very large compression ignition engines use a two-stroke cycle.

The piston engine has a combustion chamber, or cylinder which is closed at one end, while the other end is sealed by a piston that can move to and fro within the cylinder. In the spark ignition engine, a mixture of air and fuel is introduced into the cylinder through a valve at the closed end. The piston compresses this mixture inside the cylinder and the compressed mixture is ignited with a spark, producing an explosion which forces the piston to move away from the sealed end again, generating a power stroke through the motion of the piston. Once the power stroke is complete, the gases are expelled from the cylinder and the process is repeated.

In a four-stroke engine, the piston moves in and out twice for each power stroke. Starting with the piston at the top of the cylinder the first stroke involves fuel and air being drawn into the cylinder through a valve while the piston moves away from the top of the engine. When it reaches it maximum displacement it begins to return (stroke two), compressing the air/fuel mixture. At the top of the cylinder this mixture is ignited, generating the power stroke (stroke three) as the piston is forced away from the top of the engine once more. Finally, it returns again, the fourth stroke, this time expelling the spent fuel/air mixture, (the exhaust) through a second valve before the cycle is repeated. The movement of the piston is controlled by a system of levers which also convert the reciprocating motion into rotary motion of a shaft.

The two-stroke engine has a simpler cycle. The power stroke is combined with the exhaust stroke, while the inlet stroke is combined with the compression stroke. As a result, a two-stroke engine has a greater power to weight ratio because there is one power stroke to every two strokes of the engine instead of one in four. For small engines, the two-stroke cycle is relatively less efficient and can generate more exhaust gas pollutants. However, very large two-stroke engines can be highly efficient.

The compression ignition engine varies the mode of operation slightly. In this type of engine, air alone is introduced into the cylinder during the inlet stroke. This is compressed much more highly during the compression stroke that would be the case in a spark ignition engine. Compressing a gas causes it to heat up and the greater compression causes the air to become extremely hot, hot enough to spontaneously ignite a charge of fuel which is injected into the cylinder when the piston reaches the top of its compression stroke. Since ignition is achieved spontaneously, no ignition system is needed, simplifying engine design. The higher temperature in the combustion chamber (cylinder) also allows higher thermodynamic efficiency to be achieved and diesel engines are more efficient that spark ignition engines. However, the high temperature can also lead to production of greater amounts of nitrogen oxides in the exhaust gases of the engines. Compression ignition engines must be stronger than spark ignition engines in order to resist the higher cylinder pressures and this makes them more expensive.

ENGINE SIZE AND EFFICIENCY

The efficiency of a piston engine varies with the type of engine. Spark ignition engines generally have lower efficiency, with typical performance of between 28% (or less) and 42%[1] efficiency while diesel

---

[1]For spark ignition engines the efficiency varies with the amount of fuel in the fuel/air mixture. A lean mixture with less fuel produces lower emissions but is less efficient (28%). A rich mixture is more efficient (42%).

engines can achieve a maximum efficiency of 50%. Efficiency also varies with engine size; smaller engines are less efficient than larger engines. So, for example, a natural gas spark ignition engine with 100 kW electrical output might achieve an efficiency of 27%, while for a 1 MW unit the efficiency can rise to 37% and at 10 MW efficiency approaches 42%. As a consequence of this, smaller engines often have a higher exhaust gas temperature than larger engines.

From a CHP perspective, a lower efficiency is not necessarily a problem because any heat not used to generate electricity can be captured later and used for heating purposes. However, it does mean that the ratio of heat energy to electrical energy will vary with engine size. For a 100 kW engine, the ratio of heat output to electrical output may be 2:1, whereas for a large engine, the ratio will fall to less than 1:1. The balance of heat demand to electricity demand will therefore have a bearing on the size of engine when a CHP plant is being designed.

Most piston engines are available as off-the-shelf components so that the actual engine design or parameters cannot easily be varied to meet a specific requirement. The largest engines are often designed for specific applications, and in this case, it is possible to vary some of the engine operating characteristics. For example, it might be preferable to run the engine at a higher temperature, and with overall lower efficiency in order to allow more heat to be available for the heating demand.

Another consideration is engine reliability. Small piston engines are manufactured in large numbers but most are for automotive applications where engine the duty cycle is relatively light and engine lifetime is not a major consideration. Such engines are cheap but if they are used for continuous, or base load operation they are likely to prove unreliable and require regular and extensive maintenance. Larger engines are often designed for a much more demanding duty cycle, but in consequence are more expensive. There has been interest in recent years in developing the smallest engines so that they can operate continuously and reliably in order to be able to make them available for the domestic and small commercial CHP markets where electrical outputs as low as 1 kW might be needed.

With engines operating at a maximum efficiency of 50%, there will be 50% or more of the fuel input available as waste heat. In smaller engines the heat available may account for up to 75% of the energy from the fuel. When waste heat is captured and utilized, the overall efficiency of the CHP system can rise to 75% to 80%.

ENGINE HEAT SOURCES

Piston engines generate large amounts of heat during operation. Some of this heat is used to generate mechanical power which in a power generating system is used to drive a generator and produce electrical energy. However, much of it is not utilized in this way and simply emerges as waste heat. This waste heat must be captured and dumped into the environment. Otherwise the engine would overheat and fail. In consequence, engines are fitted with extensive cooling systems.

In a piston engine CHP system, these cooling systems can be exploited to provide heat energy. In most engines there are four primary sources of waste heat. These are the engine exhaust, the engine case water cooling system, the lubrication oil cooling system (where fitted) and when a turbocharger is fitted, the turbocharger cooling system. A schematic of a piston engine CHP system (without a turbocharger) is shown in Fig. 4.1.

The exhaust gas contains up to one third of the fuel energy and 30% to 50% of the total waste heat from the engine. The amount of energy it contains will depend upon the efficiency and type of engine.

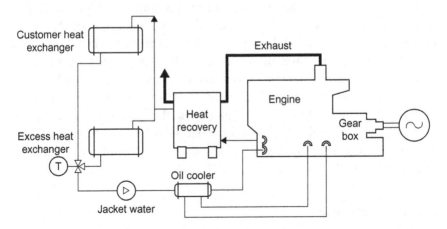

Figure 4.1 Piston engine CHP system schematic. Source: Paul Breeze (2014).

In general smaller, less efficient engines will convert less heat from the combustion gases into mechanical energy than a larger engine of the same type, so the exhaust gas temperature will be higher for the smaller engine. Diesel engines operate at a higher temperature than spark ignition engines, but they are more efficient.

Exhaust heat is not normally captured in conventional engines but it is straightforward to fit a heat recovery system to the exhaust of an engine if the heat is required. The exhaust temperature is typically between 370 and 540°C. This is sufficiently high that it can be used to generate medium pressure heat if required, with a maximum pressure of around 28 bar. The additional cost of a steam generator means that this will only be cost effective in very large engines. Otherwise, the exhaust heat can be used to generate hot water. Engine exhaust gases have also be used directly for drying in some applications.

The main engine case cooling system can capture up to 30% of the total energy input. Cooling water typically exits the cooling system at up to 95°C but it may be hotter if the cooling system is pressurized. In a cogeneration system, the outlet from the engine case cooling system will be passed through a heat exchanger to provide a source of hot water. Engine oil and turbocharger cooling systems will provide additional energy that can also be used to supply hot water. The engine cooling jacket and oil cooling system will typically provide 45% to 55% of the total waste heat recovery from an engine system.

If all the heat from the exhaust and the cooling systems of an engine is exploited, around 70%–80% of the fuel energy can be used. However this can generally only be fully exploited when there is a need for hot water. Overall efficiency will also depend on the duty cycle of the engine. Most reciprocating engines will show little fall in overall efficiency when the electrical load on the engine falls from 100% to 50% but if it falls lower than 50%, then efficiency will start to drop off more sharply. Engine and waste heat temperatures are likely to fall too, and so a wide daily variation in load is likely to have an impact on the effectiveness of the CHP system.

STIRLING ENGINE

A Stirling engine is a reciprocating engine that utilizes a piston and cylinder in the same way as a more conventional internal combustion engine. However, in the case of the Stirling engine, the system is usually pressurized and is completely sealed. Instead of combustion taking place within the engine, the device is provided with an external heat source and an external cooling system—the heat sink. Heat from the heat source passes through the engine to the heat sink, driving its cycle as it does so. Stirling engines usually contain either hydrogen or helium as their working gases because these are excellent at transferring heat quickly.

Commercial Stirling engines are relatively small. Typical electrical generating capacities are between 1 and 25 kW. While these engines potentially have a range of applications, they are often expensive to build and have rarely been used in commercial power generating systems. One area in which they have found use is for solar thermal power generation where the heat source is provided by the sun.

More recently, there has also been interest in using Stirling engines as electricity generators in domestic CHP systems. Many households, particularly in Europe, use natural gas to provide heat. A natural gas burner can easily be adapted to provide a heat source for a small Stirling engine of between 1 and 10 kW, with the size chosen to match the electrical demand of a typical domestic dwelling or small commercial organization. The engine forms part of a topping cycle system, with the waste heat that passes out of the engine through its heat sink used to provide hot water and space heating. There are a number of commercial systems based on this design available.

PISTON ENGINE CHP APPLICATIONS

The use of reciprocating engines for CHP applications is common in the developed world. In the USA, for example, there were an estimated 2000 large systems operating in the middle of the second decade of the 21st century and these had, between them, an electrical generating capacity of 2300 MW or roughly 1.15 MW for each system.[2] Large systems in the megawatt range often comprise multiple smaller engine units for added flexibility and reliability. There is likely to be a number of much smaller systems too. Similar types of system can be found in Europe and Japan.

[2]Energy Solutions Center, http://understandingchp.com/chp-applications-guide/4-chp-technologies/.

While reciprocating engines can operate on a variety of liquid and gaseous fuels, most of these CHP systems burn natural gas in spark ignition engines. Units burning gaseous fuels represent 84% of the US capacity. Natural gas is clean and relatively cheap, particularly in the USA following the development of shale gas. These systems are often packaged and can be installed in urban areas without difficulty. Typical users include universities, hospitals, office buildings as well as a range of commercial and small industrial organizations where there is a demand for both electrical power and heating or hot water. They can also provide low pressure steam if that is needed. Food processing plants represent typical users in this sector.

In addition to the use of waste heat for hot water or for generating low pressure steam, the exhaust gases from an engine can be used directly for drying of heating. One specialist application is for green-house cultivation where the extra carbon dioxide in the exhaust gases as a result of fuel combustion can boost plant growth.

The main types of reciprocating engine all have good part load efficiency, such that the efficiency will drop by 10% or less as the load falls from 100% to 50%. Modern engines for CHP applications are relatively cheap to install and are reliable provided they are maintained according to the manufacturer's schedule.

Emissions from piston engine CHP plants vary with engine type. Gas-fired spark ignition engines are generally the cleanest and most require little emission control although nitrogen oxides removal may be required. Diesel engines produce higher levels of both nitrogen oxides and particulates and will usually need exhaust gas cleanup.

The use of reciprocating engines in the domestic CHP market is new and the market for these units is developing slowly. Initially, the main markets are likely to be in Japan and in Europe.

CHAPTER 5

Steam Turbine Combined Heat and Power Systems

Steam turbines are heat engines that extract energy from high temperature, high pressure steam and convert it into rotational, mechanical energy, driving a shaft that can be used to turn an electrical generator and provide electricity. Unlike some heat engines, a steam turbine does not, itself, burn fuel. In order to operate, a steam turbine needs a source of high pressure, high temperature steam. This is normally provided by an integrated furnace and boiler that is designed to match the steam turbine. In some cases the heat energy for steam generation may also be supplied from the exhaust of a gas turbine or as waste heat from another process. In consequence, a steam turbine-based electrical generating or combined heat and power (CHP) system is relatively complex and expensive. Against this, steam turbine CHP systems can provide high quality heat for industrial processes with great efficiency and flexibility.

Steam turbines are the mainstay of the world's electrical generating capacity, providing mechanical power to drive generators in coal-fired, oil-fired, some gas-fired and nuclear power plants across the globe. They can also be found in geothermal power plants, in some solar thermal power plants and in marine power plants that derive heat from the world's oceans. The steam turbines in large power plants such as central coal-fired and nuclear plants are usually made up of several cascaded units that extract energy in turn from high pressure steam, medium pressure steam and low pressure steam. Smaller plants will have a single steam turbine and so will most steam turbine based CHP plants, although some of the very largest plants might include cascaded units.

Steam turbine CHP systems can be found in a variety of configurations with steam provided from boilers burning a variety of fuels. Some will burn coal but many will be fired with natural gas. In industry-specific situations, the fuel may be biomass or some form

Combined Heat and Power. DOI: https://doi.org/10.1016/B978-0-12-812908-1.00005-5

of waste material. Many of these installations will be topping cycle CHP systems but steam turbines can also be used in bottoming cycles. Topping cycle steam turbine CHP systems will normally be designed to provide process heat in the form of steam for some industrial or commercial process. This may require steam at a variety of temperatures and pressures and CHP steam turbines are designed to be able to provide steam at different conditions to suit the demand. Similar steam turbines can be used in district heating systems.

These is another type of turbine called an organic Rankine cycle (ORC) turbine that operates in a manner identical to that of a steam turbine but uses an organic fluid as its working fluid instead of steam/water in a steam turbine. These devices are usually closed cycle units that can exploit relatively low temperature waste heat sources. As such they can be used in CHP bottoming cycles when the waste heat is not of sufficient quality to raise steam.

STEAM TURBINE TECHNOLOGY AND TYPES

The steam turbine harnesses the principle that a high pressure jet of fluid can be used to impart rotational drive to a wheel fitted with paddles or blades. This is the principle of the water wheel. In the case of a steam turbine, the fluid in question is high pressure, high temperature steam.

The water wheel operates most efficiently when its blades move at half the speed of the flowing water. High pressure steam exiting a nozzle can reach 1500 m/s, and it would be impossible to harness this with a simple bladed wheel because it could never rotate fast enough to operate efficiently. The solution is to design a steam turbine with a series of wheels and blades, with the steam temperature and pressure dropping in stages as it passes each one. In this way, the energy can be extracted efficiently without the need for impossibly high blade speeds.

A modern steam turbine comprises a series of bladed wheels mounted sequentially along a shaft. The blades and shaft are contained inside a casing through which the steam flows. In between each set or rotating blades are a set of fixed blades. These act as nozzles, allowing the steam to expand and increase in velocity sequentially, while the rotating blades exploit the kinetic energy of the steam at each stage. In this way, the steam pressure and temperature decreases as it passes through the steam turbine. The thermodynamic heat engine cycle upon which the steam turbine is based is called the Rankine cycle.

A steam turbine will extract the greatest amount of energy from the steam when the pressure and temperature drop across the sets of blades is a maximum. In a conventional power plant, this is achieved by condensing the steam at the exit of the steam turbine. A CHP plant may require heat energy to be left in the steam when it leaves the steam turbine. In this case, not all the energy can be taken from the steam to generate electrical power. Alternatively, steam may be extracted from a steam turbine at an intermediate point along its length where the temperature and pressure of the steam within the turbine matches that required by heat demanding process.

Since the efficiency of a steam turbine power plant depends on the temperature and pressure of the steam, modern technology attempts to maximize this by raising the temperature and pressure above the critical point of water. Beyond this point, liquid and gaseous water are indistinguishable. Supercritical steam plants of this type can operate with a steam temperature of around 600°C and a pressure close to 300 bar. At operating temperatures in this region, both boiler and turbine components must be made of special, expensive materials and this will not normally be cost effective in a CHP installation. Most will therefore operate with steam below the supercritical point.

CHP STEAM TURBINE TYPES

In order to match steam production to heat demand, there are three types of steam turbine in common use for CHP plants. These three are called a back pressure steam turbine, an extraction steam turbine and a condensing steam turbine. Each type will suit a different configuration of plant.

The condensing steam turbine is a conventional steam turbine in which the steam exiting the turbine exhaust is condensed to provide the maximum energy capture. This is the type of steam turbine that will be used in a bottoming cycle CHP plant in which waste heat or energy from an industrial process is used to raise steam to generate electricity. In this type of plant heat energy is exploited first, then the remaining heat is used to produce steam for electricity generation.

The other two types of steam turbine are more commonly found in topping cycle systems. The back pressure steam turbine is so-called because the steam exiting the turbine exhaust will still contain a significant amount of energy and will be at relatively high temperature and pressure. This exhaust steam is used either to provide heat for an industrial process or to produce hot water. A back pressure turbine is shown schematically in Fig. 5.1. This type of steam turbine CHP plant will normally operate with relatively constant heat demand because there is no way of varying the steam output from the turbine exhaust while maintaining optimum operating conditions.

The third type, an extraction steam turbine, is a condensing steam turbine with a steam extraction port part-way along the steam turbine that can provide steam at intermediate temperature and pressure for an industrial process (see Fig. 5.2). The actual position of the extraction port will depend on the heat demand so this type of CHP plant will be tailored for each specific application. In an extraction steam turbine CHP plant the amount of power that can be generated will vary, depending upon the amount of steam that is extracted. This makes for a flexible plant in which power and heat can be balanced depending upon the heat demand. However, the turbine is more costly that a back pressure or conventional condensing steam turbine.

In addition to these three types, there is a hybrid extraction/back pressure steam turbine that provides steam for process heat or hot water both from an extraction point in the turbine and from the turbine

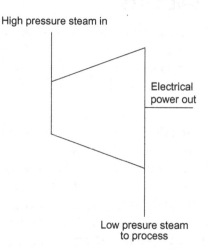

Figure 5.1 Schematic of a back-pressure steam turbine.

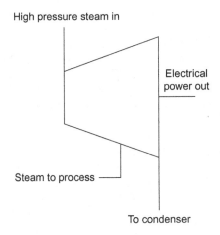

High pressure steam in

Electrical
power out

Steam to process

To condenser

Figure 5.2 Schematic of an extraction steam turbine.

exhaust. This type of CHP plant is the most complex to control because the heat flow through the turbine will depend upon the amount of heat extracted, and this will cause the heat at the turbine exhaust to vary.

Steam turbines are available in a wide range of sizes making it relatively simple to match a turbine to a particular CHP application. The complexity of steam turbine CHP plants means that they are normally relatively large in size, and the smallest CHP steam turbines are likely to be several megawatts in generating capacity. Where heat demand is smaller, alternative technologies are more commonly used. Maximum size for a single steam turbine is around 250 to 300 MW.

PLANT CONFIGURATIONS

In addition to the variety of steam turbines available for CHP applications, there are a range of plant configurations, depending on the heat source, and whether the turbine is to be used in a topping cycle or a bottoming cycle. Energy sources can include fossil fuels and biomass. Solar thermal energy can also be harnessed to raise steam to drive a steam turbine but this is unlikely to be incorporated into a CHP plant.

A large modern supercritical coal fired power plant will probably be capable of converting up to 47% of the energy in the fuel into electrical energy although many plants operate at much lower efficiency. When a conventional plant of this type is used for CHP, the electrical efficiency

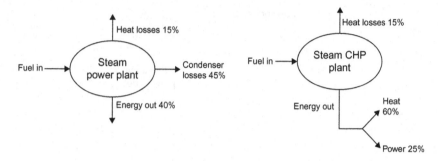

Figure 5.3 Schematics of conventional power plants with and without CHP.

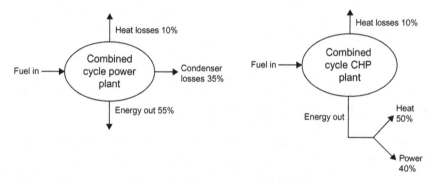

Figure 5.4 Schematics of combined cycle power plants with and without CHP.

may drop as low as 25% but a further 60% of the energy from the fuel can be extracted and used as heat, providing an overall efficiency of up to 85%. This is shown schematically in Fig. 5.3.

An alternative to the coal-fired power station is a natural gas fired power plant. Today, this is likely to be a combined cycle power plant in which natural gas is burned in a gas turbine to generate electricity and then the heat energy in the gas turbine exhaust is used to raise steam, driving a steam turbine to produce more power. The best combined cycle power plants of this type can achieve an efficiency of 60% of slightly higher. Smaller plants tend to have lower efficiencies. It is possible to adapt this type of plant to CHP by extracting heat from the gas turbine exhaust, normally by generating steam, but the exhaust gases can be used directly as well if the process requires it. Depending on the balance between electricity and heat, a plant of this type might achieve an electrical efficiency of up to 40% with a further 50% of the energy being captured as heat (see Fig. 5.4).

More complex configurations include the addition of supplementary boilers or the introduction of supplementary burners in the heat recovery steam generators. This allows steam production to be boosted. Configurations of this type have been used in refinery CHP plants.

ORGANIC RANKINE CYCLE TURBINES

An ORC turbine exploits exactly the same thermodynamic principle (the Rankine cycle) as a steam turbine with the single difference that the working fluid used by the turbine is not water/steam but an organic compound with a relatively low boiling point. The low boiling point allows the liquid to be vaporized at a much lower temperature than water requires to convert it into steam and this allows a lower temperature heat source to be exploited.

Fluids used in ORC are often refrigerants or hydrocarbons. Suitable compounds need to be stable at the temperature at which the heat is being recovered and have a freezing point that is lower than the lowest temperature reached in the cycle. Liquids with a high latent heat of vaporization are preferred because they absorb more heat for each unit of liquid and this minimizes the overall system size.

An ORC system will have a heat exchanger through which the fluid carrying the waste heat is passed so that the heat can be used to vaporize the ORC fluid, a turbine to extract the energy from the vaporized fluid and a condenser to return the fluid to the liquid state after the turbine. All these components will form a closed system so that none of the fluid can escape. A simplified schematic of an ORC system is shown in Fig. 5.5. ORC turbine systems will normally come as packages ready to be linked to the heat source and the power supply. System size is relatively small, typically from 100 kW to 1 or 2 MW although systems as large as 20 MW are available.

Typical applications for this type of turbine include extraction of energy from low temperature geothermal sources, small-scale biomass plants, domestic CHP and waste heat recovery using a bottoming cycle from a range of different sources, and solar thermal power generation. The ability to exploit low temperature waste heat sources means that these turbines are particularly suitable for CHP bottoming cycle applications.

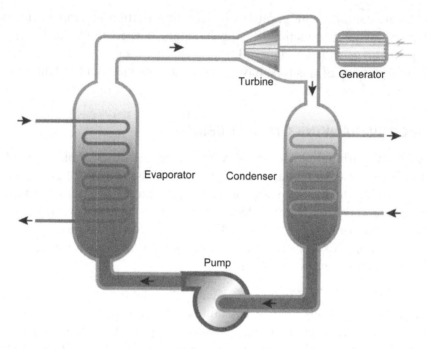

Figure 5.5 Schematic of an organic Rankine cycle system. Source: Calnetix.com.

EMISSIONS

The atmospheric emissions from a steam-turbine-based CHP plant will depend upon the fuel type and plant configuration. Many CHP plants of this type will burn natural gas which is relatively clean although there will be a need for exhaust gas cleanup to remove nitrogen oxides, carbon monoxide and particulates. If gas turbines are used to provide both electrical power and heat, these will need to be fitted with control systems similar to those in more conventional gas turbine power plants. Coal burning plants generally require much more extensive exhaust gas cleanup since the fuel is much dirtier. The measures needed for other fuel types will depend upon that fuel. The exhaust gases from biomass fuels usually require some clean cleaning, but less than for a coal-fired plant.

STEAM TURBINE CHP APPLICATIONS

Steam turbine CHP plants are most widely used in the industrial sector where high quantity and high quality heat is required. In general these plants will be designed around the heat demand with electricity as the secondary product. Industries that commonly use steam turbine-based CHP plants include the ammonia and fertilizer industries, chemical plants, pharmaceutical plants, the pulp and paper industries, sugar and food plants, and power-and-desalination plants. According to Cogen Europe, the paper and allied industries account for 20% of industrial CHP installations, the chemical and allied industries for 40% and the petroleum and refining industries for 15%.

The CHP plants in some of these industries can be complex. For example a refinery CHP plant in the United Kingdom that was completed in 2004 comprises two 260 MW gas turbines with heat recovery steam generators, each equipped with supplementary firing, and two auxiliary boilers together with two 115 MW steam turbines which proved steam at 50 bar for refinery use. Depending upon heat demand the plant can generate between 150 and 734 MW of power which is exported to the grid.[1]

Other industries require steam at different pressure and temperature. The paper making industry typically needs steam at 3.5 bar and 145°C. Wood processing and paper making plants often have CHP plants that burn waste from the processes to generate steam and power that is used by the plant. For the evaporation of juices in sugar making, saturated steam at around 3.5 bar is also used. Saturation of the steam increases its heat transfer properties.

Large district heating plants are likely to be based on steam turbines because these offer the highest efficiency in many advanced generating plants. Modern CHP plants for district heating can use complex configurations to achieve this high efficiency. A plant in Dusseldorf uses a gas turbine combined cycle configuration in which 300 MW of thermal energy are extracted from the low pressure steam turbine to provide energy for the district heating load. This plant has a claimed CHP efficiency of 85%.[2]

---

[1]Inaugurating Immingham: Europe's biggest CHP, Modern Power Systems, November 2004.
[2]Development trends in cogeneration and combined heat and power plants, Andreas Pickard and Frank Strobelt, Siemens, Power-Gen Europe, 2016.

CHAPTER 6

Gas Turbine Combined Heat and Power Systems

Gas turbines are heat engines that use air as their working fluid. The thermodynamic cycle upon which these turbines operate is called the Brayton cycle. It differs from that of the steam turbine in some details but the two cycles share many common features. Where the two types of machine diverge is that a modern gas turbine combines an air compressor, a combustion chamber and a power turbine in a single device with the compressor and turbine blades all mounted onto a single shaft. The gas turbine is therefore a completely self-contained unit that requires only fuel and air in order to generate mechanical, rotational power. The steam turbine, in contrast, requires a separate furnace and boiler to function.

The modern gas turbine was developed in the early and mid-20th century, initially as an aero engine, before being adapted for power generation. The first turbines used by the power industry were closely modeled on their aero-engine cousins. These aero-derivative gas turbines are still available today and offer the highest efficiency of any gas turbines.

Aero engines need to be very light and very efficient. There is no similar weight limitation for gas turbine engines intended for stationary applications such as power generation. This has led to the evolution of a separate class of heavy duty, or industrial gas turbines that are designed specifically for this market.

Until the 1980s, the majority of gas turbines in power plants operated in what is known as open cycle mode. Since then it has become popular to use them in combined cycle power plants in which heat from the gas turbine exhaust is captured and used to raise steam to drive a steam turbine. Plants of this design can reach efficiencies of 62%.

Most gas turbine power plants burn natural gas although they can usually be fired with liquid fuels too. Natural gas is popular for power

Combined Heat and Power. DOI: https://doi.org/10.1016/B978-0-12-812908-1.00006-7

generation because it is clean compared to coal and the high efficiency of combined cycle plants means that the emissions of carbon dioxide are lower than from a coal plant. Gas turbine-based power plants are also much cheaper to build than coal-fired power stations. However coal is historically much cheaper than natural gas. One consequence of this different cost structure is that the economics of gas turbine power plants depend to a large extent on the cost of natural gas. In the middle of the second decade of the 21st century natural gas is relatively cheap, particularly in the USA. However there have been periods in the recent past where regional gas prices has spiked, making gas turbine power plants uneconomical to run.

In combined heat and power (CHP) applications, a gas turbine will almost always be used in a topping cycle configuration with heat being recovered from the exhaust of the engine and used in an industrial process or to provide heating and hot water. There are a small number of situations in which a gas turbine could potentially be used in a bottoming cycle. These arise when there is a stream of high pressure, high temperature air that is exiting from an industrial process. A power turbine can extract energy from this gas stream.

Conventional gas turbines are available in sizes between around 1 and up to 400 MW. All are high technology products that are manufactured by a limited number of companies, globally. In addition to these relatively large gas turbines there are a number of much smaller microturbines available. Microturbines are technically simpler gas turbines with capacities of from a few kilowatts to a few hundred kilowatts. Many of these are aimed at domestic or small commercial power and CHP applications.

GAS TURBINE TECHNOLOGY

A commercial gas turbine for power generation and CHP applications consists of three components, an axial compressor, a combustion chamber and a power turbine. These components are integrated into a single unit and with the blades of axial compressor and those of the power turbine often mounted onto the same shaft. This shaft is coupled to a generator in order for the system to produce electric power. Compressor and turbine are housed in a casing or shroud that contains the hot, high pressure gases and constrains them to flow through the compressor and turbine.

In operation, air is drawn into the compressor intake where the spinning blades compress the gas. The compressed air is then fed into a combustion chamber where it is mixed with fuel and ignited, producing a very hot, high pressure stream of air and combustion products. This stream of gas is directed into the power turbine section which is equipped with several sets of turbine blades, interspersed with fixed blades or nozzles similar to those in the steam turbine. The gases flowing through the turbine generate power in the form of rotational motion of the engine shaft as energy is extracted from the gas stream. Finally, the cooler gases are exhausted from the turbine where they can be exploited for heating purposes.

The energy produced by the power turbine stage is sufficient both to drive the compressor and provide the additional mechanical energy needed to turn the generator. The best optimized aero-derivative gas turbines can achieve an input energy to electrical conversion efficiency of 46%. Units of this type are usually under 100 MW in generating capacity. Much smaller aero-derivative gas turbines will have lower efficiencies. Industrial or heavy duty gas turbines cannot achieve this level of efficiency but the best will have an efficiency of between 38% and 42%. Lower efficiency is not usually a problem with these turbines because most are used in combined cycle plants where waste heat is captured for steam production. The same applies when a turbine is used for a CHP application.

The temperature of the exhaust gases exiting a high efficiency aero-derivative gas turbine will be in the range 400 to 500°C. With a lower efficiency heavy duty gas turbine, it could be higher. These hot gases can be used to produce steam in a heat recovery steam generator (HRSG) and that steam can be used to provide heat for an industrial process, for district heating or to drive a steam generator. (The hot gases can also be used directly to heat a kiln or for drying.) Additional steam can be produced by adding supplementary burners inside the steam generator, or by the use of an auxiliary steam generator. Some CHP plants combine heat provision with additional electricity generation from a steam turbine. Such plants are relatively complex but they are also very flexible. Large plants of this type can provide significant amounts of power to the grid as well as industrial heat or energy for district heating.

Although gas turbines are complex, technically advanced machines they are relatively cheap compared to other prime movers for power generation. In addition the complete unit is built in a factory before being shipped to the site, eliminating much of the on-site fabrication that may take place with other types of power station. This has made them a popular choice for both power generation and for CHP plants across the world. Many of the CHP applications supply heat and power for large industrial sites or large district heating plants.

GAS TURBINE CHP PLANT CONFIGURATIONS

Gas turbines are usually part of an electricity generation topping cycle. The simplest configuration for a gas turbine CHP plant is when the gas turbine is used to provide electrical power while the hot exhaust gases are used directly for drying or in a kiln. In some cases it is possible to use the exhaust gases directly in a kiln, in others air-to-air heat exchangers are used to provide hot air. Auxiliary firing can be added to provide additional hot air if needed.

The most common configuration, however, involves the gas turbine acting as a topping unit, producing electrical power, while heat from the exhaust of the machine is used to generate steam or hot water. A simple schematic of this type of gas turbine CHP plant is shown in Fig. 6.1. In plants using this scheme, the balance between electricity and heat demand will determine the exact configuration of the plant.

In many industrial CHP applications, there will be a need for both power and heat. However the two are rarely balanced so well that a single gas turbine can provide for both. More often, the gas turbine will be chosen to meet the local electricity demand but the heat demand will be greater than a HRSG fitted to the exhaust of the turbine can supply. The solution in this case is to add supplementary firing burners inside the HRSG to provide additional steam raising capability.

For an installation providing steam at 12 bar and 200°C, typical of industrial CHP demand, the amount of heat available will depend on turbine size and efficiency. For example a 12.5 MW gas turbine may be able to provide up to 25 tonnes/h of steam without supplementary firing while a 45-MW gas turbine can provide around 80 tonnes/h.[1] With supplementary firing the amount of steam available from each can be doubled.

[1]Gas Turbine Cogeneration Activities—New Power Plants for Uralkali's Facilities, Alexander Gushchin, Ian Amos and Guy Osborne, Siemens.

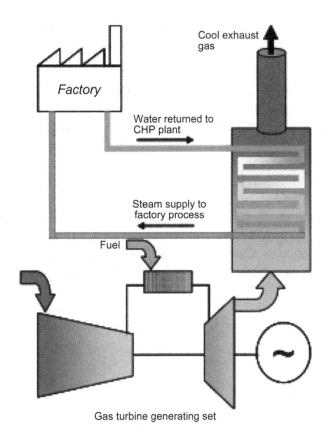

Figure 6.1 Schematic of a gas turbine CHP plant. Source: Siemens[2].

An alternative approach is to select the size of the gas turbine so that the heat in its exhaust gases will provide all the local heat demand. This will often result in there being a surplus of electricity. This additional power can be exported to the local grid provided this is permitted by national regulations. It is usually possible in most countries around the world today.

As well a heat, a CHP plant can provide cooling too. This is the basis for a trigeneration plant in which a gas turbine is used to generate electrical power and then part of the steam generated by heat captured in a HRSG is used for an industrial process or for heating while the remainder of the steam is used to drive an absorption chiller. In a

[2]Alexander Gushchin, Siemens Russia Ian Amos, Product Strategy Manager, SGT-400, Siemens Industrial Turbomachinery Ltd, UK. Guy Osborne, Sales Manager, Siemens Industrial Turbomachinery Ltd, UK.

plant of this type, the balance between heating and cooling can be varied according to the local need. Most cooling will be for air-conditioning and so demand may well vary with the season.

A more complex type of gas turbine CHP plant is the gas turbine combined cycle CHP configuration. In a standard combined cycle power plant a gas turbine is coupled with a HRSG and a steam turbine in a highly integrated plant with both gas and steam turbines driving one of more generators to provide electrical power. Fully optimized plants of this type can achieve up to 62% efficiency. This configuration can readily be adapted to CHP with steam extracted either directly from the HRSG or from an extraction steam turbine (see Chapter 5) to provide heat. A schematic of a gas turbine combined cycle CHP plant is shown in Fig. 6.2. Balancing a plant of this type will be challenging because the steam flow through the steam turbine

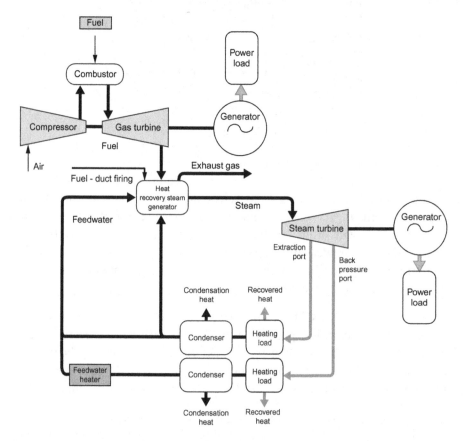

Figure 6.2 Schematic of a gas turbine combined cycle CHP plant. Source: www.Retscreen.net.

will depend on heat demand and as this varies, so will the operating conditions of the steam turbine. Plants of this configuration have been used in refineries and for district heating. Such plants are most often grid connected so that most or all of the electricity produced by the plant is exported to the grid.

An important variant of this type of CHP plant is the desalination plant designed to produce drinking water from sea water. Combined cycle desalination plants have become particularly popular in the Middle East where there is a large demand for potable water, with limited local supply, coupled with a ready supply of natural gas. Desalination is usually carried out thermally with steam produced from a gas turbine exhaust. In a typical plant, steam for thermal desalination is extracted from the plant steam turbine although it may also be taken directly from the HRSG. For example, a sea water combined cycle desalination plant in Kuwait is fitted with a 300 MW extraction steam turbine from which steam is taken to supply two multistage flash desalination plants. Depending on demand from these units, the electrical output of the steam turbine varies from 300 MW down to 75 MW.

It is possible, in principle, for a gas turbine to be used in a bottoming cycle although CHP applications of this type are likely to be rare. In order for this to be possible a stream of relatively high temperature, high pressure air is required. Simple air turbines (these are essentially the power turbine of a gas turbine without the compressor or combustion chamber) have been considered at a means to extracting additional energy from the exhaust of a gas turbine, or even a piston engine, but this is essentially a combined cycle configuration. There are industrial processes that produce suitable gas stream such as high temperature furnaces but the energy they contain is more normally captured using a steam turbine or organic Rankine cycle system. However, there is interest in the development of air turbines for this purpose.

MICROTURBINE CHP

A microturbine is a small gas turbine designed for use in domestic and commercial installations. Identical in concept to standard gas turbines, these devices are much simplified with perhaps a single set of compressor blades and a single set of power turbine blades. Microturbines can be designed either for power generation alone or as CHP units. Although they are capable of burning a variety of fuels, most will be intended to operate with natural gas. The units are usually designed as a complete package for electricity and hot water production. All that is required is to connect the package to the electricity supply, a gas supply and to the hot water system.

Microturbines are available in sizes ranging from 30 kW up to 400 kW. Beyond that conventional gas turbines take over. There are, in addition, much smaller microturbines aimed at the domestic market. These have electrical generating capacities of 1 kW to 10 kW. All microturbines operate at extremely high speed, with rotational speeds often in excess of 60,000 rpm. The smaller the turbine, the higher the speed. Electrical efficiency of these small machines is relatively low with a 30-kW machine typically capable of around 23% efficiency. Larger machines are slightly more efficient. CHP efficiency is much higher and a 30-kW microturbine might achieve 67% CHP efficiency.

GAS TURBINE EMISSIONS

The atmospheric emissions from gas turbine are relatively low if they are fired with natural gas. They can be slightly higher with liquid fuels, but all gas turbines are sensitive to fuel impurities which can damage their components so these fuels are generally cleaned and emissions are therefore limited. The most important emissions are nitrogen oxides which, at high levels, can cause a range of health and environmental problems. The combustors in modern gas turbines are designed to minimize the production of these gases but the emissions from large gas turbine CHP plants will normally require some form of nitrogen oxide emission control. The most common is selective catalytic reduction in which a reagent such as ammonia is mixed with the exhaust gases and then reacted over a catalyst to reduce the nitrogen oxides back to nitrogen.

Other emissions that can be generated include carbon monoxide and unburnt hydrocarbons. Both of these can be controlled using an oxidation catalyst. Microturbines are usually designed so that their emissions will be below local regulatory limits, and they can be installed in urban settings without a problem.

Since gas turbines burn fossil fuels, they will also generate carbon dioxide from the combustion of the fuel. Some turbine plants operate on biogas that is environmentally neutral but most contribute to the atmospheric burden of carbon dioxide.

GAS TURBINE CHP APPLICATIONS

Gas turbine CHP plants are capable of serving a wide range of heat and power needs. Some of the largest plants operate as grid power generators while providing either process heat for industrial heat users or hot water for district heating. Plants of this type may be gas turbine cogeneration or gas turbine combined cycle cogeneration plants, and they usually operate with high efficiency, often up to 85% in CHP mode. Plant electrical generating capacities can be several hundred megawatts. In the Middle East, combined cycle cogeneration plants are popular as desalination facilities.

Smaller CHP plants based on small industrial or aero-derivative gas turbines are commonly used by industries that have a need for both heat and power. These plants usually have a limited electric power capability, with output sufficient only for the industry they serve, and frequently use supplementary firing to provide extra steam or heat output.

Large microturbines might also be used in an industrial setting but most of these units are designed to provide power, heating and hot water. Typical applications include hospitals, retail centers, office blocks and leisure and recreation centers. Small microturbines, in the 1 to 10 kW range, are aimed at the domestic and small commercial market. This is a relatively new area for CHP units of any type and there are many competing technologies including Stirling engines and fuel cells. Many of the companies producing these CHP units are targeting domestic markets in Europe and Japan.

CHAPTER 7

Fuel Cell Combined Heat and Power

The fuel cell is an electrochemical device, similar in concept to a battery, that exploits a spontaneous chemical reaction to produce electricity. Chemical reactions that proceed spontaneously are called exothermic reactions because these reactions release heat. By manipulating the way in which such a reaction takes place, it is possible to convert much of the heat energy into electrical energy and this forms the basis for the electrochemical cell.

While the energy available for electricity production can be equated with heat energy from the reaction, batteries and fuel cells are not heat engines and they are not bound by the same thermodynamic limitations as heat engines. This means that it is possible to extract much higher efficiency from a device of this type than from a conventional heat engine.[1] The theoretical maximum efficiency of a fuel cell at room temperature is 83%. While most real cells will not come close to this, some practical fuel cells can reach efficiencies of 60%.

Conventional batteries exploit a wide range of different chemical reactions but most fuel cells rely on only one, the reaction between hydrogen and oxygen to produce water. This makes the fuel cell a conceptually simple device. The reaction between these gases is highly exothermic. However it does not take place if the gases are mixed at ambient temperature; initiation of the reaction requires either a high temperature, a spark or a catalyst to set it off. However once it has started the heat it produces is sufficient to cause a chain reaction to take place. Controlling the reaction is, therefore, a key to fuel cell operation. There are a range of fuel cell designs that exploit the reaction, some operating at or close to room temperature, some at extremely high temperature.

The use of a fuel cell as a combined heat and power (CHP) device depends on some of the energy from the reaction emerging as heat. Depending upon the exact cell design, it may be possible to capture

---

[1]This is true at room temperature. At high temperatures the heat engine can be more efficient.

Combined Heat and Power. DOI: https://doi.org/10.1016/B978-0-12-812908-1.00007-9

this heat and used it for heating and hot water, or as a heat source for an industrial process. There are some fuel cells that have been developed specifically with CHP in mind. In the case of others the potential for CHP exists but has not been exploited. There has been a great deal of interest and investment in developing the fuel cell as an automotive power source. This has helped accelerate the development of some fuel cell technologies. There is also interest in using fuel cells in small domestic CHP units.

Commercial fuel cell CHP systems are predominantly aimed at providing heating and hot water in addition to electrical power. Production of low pressure steam is also possible. The fuel cell is a very clean, quiet power generation system and is ideal for use in urban installations.

FUEL CELL TECHNOLOGY

The fuel cell is an electrochemical cell which delivers electrical energy from a chemical reaction. It is similar to a battery, but while most batteries carry the chemical reactants around with them, the chemicals required by the fuel cell to generate power are fed to it from external sources. This means that so long as there is a supply of chemical fuel the fuel cell can operate. In contrast a battery will become exhausted once the chemical reactants within it are used up.

Virtually all fuel cells that have been developed commercially use only two reactants, hydrogen and oxygen, the latter normally supplied from air. There is one major exception, the methanol fuel cell, which can use methanol instead of hydrogen. Hydrogen is not widely available, and most commercial cells rely instead on natural gas which is first "reformed" to convert it into hydrogen.

Like all electrochemical devices, the fuel cell operates by separating the reaction into two parts, one taking place at each electrode of the cell. These two electrodes are separated by an electrolyte. This electrolyte will allow charged atoms to pass from one electrode to the other but it will not conduct electrons. However electrons also need to pass from one electrode to the other if the reaction is to be completed. In order to do so, they must pass through an external circuit and this is the source of electrical power. A block diagram of a fuel cell is shown in Fig. 7.1.

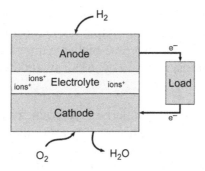

Figure 7.1 Fuel cell block diagram. Source: Wikipedia.

The reaction between hydrogen and oxygen is simple:

$$H_2 + O_2 = H_2O$$

In the fuel cell, hydrogen is delivered to one electrode, called the anode, and oxygen or air to the second, called the cathode. When hydrogen is fed to one electrode of the cell and oxygen to the other, the reaction is broken into two half reactions:

$$\text{Anode reaction:} \quad 2H_2 = 4H$$
$$4H = 4H^+ + 4e^-$$

$$\text{Cathode reaction:} \quad O_2 = 2O$$
$$2O + 4e^- = 2O^{2-}$$

$$\text{Cell reaction:} \quad 4H^+ + 2O^{2-} = 2H_2O$$

Depending upon the type of electrolyte, either the charged hydrogen ions generated at the anode or the charged oxygen ions from the cathode reaction will be able to pass through the electrolyte while the electrons that are required to complete these reactions pass through an external circuit. Fuel cell types are characterized by the type of electrolyte they use.

Hydrogen and oxygen are normally in molecular form when provided to a fuel cell electrode and the molecules do not dissociate readily, except at very high temperatures. However the cell reactions require that they are present at each electrode as atoms. This is accomplished by using catalysts that promote the molecules to dissociate. These catalysts are usually metals or metal oxides. For low temperature cells, the most important catalyst is platinum. High temperature cells often use nickel or nickel oxide.

If a fuel cell is fed with pure hydrogen and oxygen from air, it can be extremely efficient and some simple low temperature cells can reach 60% efficiency. Efficiency drops at higher temperatures and the maximum theoretical efficiency is only 62% at 1000°C. The lost energy emerges as heat.

Most modern cells have to derive hydrogen from natural gas using a process called reforming. This is an energy intensive process. It can be driven using heat energy released by the fuel cell reaction (or by burning some natural gas), but this reduces the overall cell efficiency. In a CHP plant, most fuel cells are capable of achieving in excess of 80% overall efficiency.

A single fuel cell provides an output voltage of between 0.5 and 0.8 V with the current capability depending upon the cell area. To achieve useful voltages, fuel cells are stacked in series. These stacks require cooling to remove the excess heat released during the cell reaction. The cooling circuit is one of the main sources of CHP heat, along with the heat carried away in the exhaust gas streams from the electrodes. For a low temperature fuel cell, the ratio of heat energy to electrical energy is roughly 2:1.

CHP AND FUEL CELL TYPES

There are five types of fuel cell that have been developed commercially and all can potentially be used in CHP systems. The five types are the proton exchange membrane fuel cell (PEMFC), the phosphoric acid fuel cell (PAFC), the molten carbonate fuel cell (MCFC), the solid oxide fuel cell (SOFC) and the alkaline fuel cell. The main characteristics of these cells are shown in Table 7.1.

In addition to these five, there is a relative newcomer, the direct methanol fuel cell. However, this is being developed to supply power for small portable devices, and CHP units based around this cell have not yet been proposed. The basic CHP configuration for each fuel cell type is similar with heat being captured from the exhaust gases of the plant and used for heating or steam production. A schematic of a generic fuel cell CHP plant is shown in Fig. 7.2.

The alkaline fuel cell. The alkaline fuel cell is the first, and in some ways, the most famous of fuel cells since it was used in the US space

Table 7.1 Characteristics of Main Fuel Cell Types			
Fuel Cell	Typical Operating Temperature (°C)	Typical Size (kW)	Efficiency
Alkaline fuel cell	<100	1−100	>60%
Proton exchange membrane fuel cell	<120	1−100	60% with hydrogen
			40% with reformed fuel
Phosphoric acid fuel cell	150−200	5−400	40%
Molten carbonate fuel cell	600−700	300−3000	50%
Solid oxide fuel cell	500−1000	1−1000	60%
Source: US Department of Energy, Fuel Cell Today.			

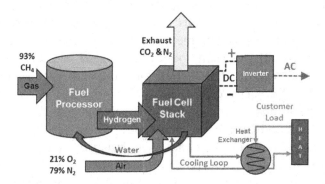

Figure 7.2 Schematic of a fuel cell CHP system. Source: Energy Solutions Center.

program that took men to the moon, as well as powering the space shuttle. This cell uses an electrolyte that is composed of concentrated alkaline solution, usually potassium hydroxide. The cell reactions in this cell are slightly modified because the electrolyte conducts hydroxyl ions (OH^-) rather than oxygen ions, but the overall reaction is essentially the same as described above. Cells can be operated at up to 260°C, and at this temperature, a nickel catalyst will suffice. However, modern cells often operate at much lower temperatures and require a precious metal catalyst such as platinum. These cells are among the most efficient to be developed and have shown efficiencies in excess of 60%. However, they need a purified source of hydrogen to operate and are sensitive to carbon dioxide in the fuel which can poison the electrolyte. Lifetime for these cells is generally lower than for other types and extending the lifetime has been a focus for much recent research. Alkaline fuel cells are being developed to form the power generating element of a domestic CHP system with waste heat used for hot water and heating. The fuel cell stack in these domestic systems are operated at a relatively high temperature to avoid the use of expensive catalyst materials.

The proton exchange membrane fuel cell. The PEMFC has an electrolyte that is formed from an acidic polymer similar in structure to Teflon but with acidic molecular groups grafted onto it. When the polymer is hydrated by saturating it with water, it will conduct hydrogen ions. The polymeric electrode material is easy to fabricate in thin sheets and electrodes and catalysts are applied to its surfaces using similar techniques to those for microchip fabrication. The presence of water means that the cell temperature must be kept below 80°C although some pressurized cells operate at higher temperature. This means they require precious metal catalysts to operate. To keep the fuel cell stack stable, efficient cooling is required, and this can provide hot water and heating. The lifetime of these cells is typically around 10,000 h. The simplicity of the PEMFC has made it a leading candidate for the power source in fuel cell vehicles and motor company investment has advanced the technology more rapidly than that of some of the other fuel cell technologies. Aside from automotive applications, the most important potential use for these cells is in domestic CHP systems. Japan has been supporting their development and units of around 750 W are being installed in Japanese homes.

The phosphoric acid fuel cell. The PAFC utilizes an electrolyte made from phosphoric acid which is a solid at room temperature but melts at 42°C. The acid begins to decompose beyond 200°C so the operating temperature is held just below this. The electrolyte is a thin layer held within a solid matrix with electrodes attached to the two sides of the matrix. Each electrode is coated with a platinum-based catalyst to accelerate the cell reactions. It is important with this cell to ensure that water formed during the cell reaction is removed from the cathode, otherwise the cell may become waterlogged. The PAFC was the first fuel cell to be marketed commercially, in the form of 200 kW CHP units, and it remains one of the most important fuel cell technologies. Typical electrical efficiency for a commercial unit is 36% but with heat recovery this rises to between 70% and 80%. Around half of the heat is recovered from the stack cooling system and the remainder from the exhaust gas streams from the electrodes. Most PAFC fuel cell units are designed to provide CHP for small commercial or municipal organizations such as hospitals. Typical unit sizes are between 100 and 400 kW. PAFCs have lifetimes of up to 40,000 h making them well suited to stationary operation.

The molten carbonate fuel cell. The MCFC is a high temperature fuel cell which operates at around 650°C. The electrolyte for this cell is a mixture of lithium carbonate and potassium carbonate which melts at 550°C. The cell reaction is the most complex of all fuel cells because it involves conduction of the carbonate ion (CO_3^{2-}) through the molten carbonate to complete the cell reaction. (Even so, the overall reaction is that between hydrogen and oxygen to produce water.) The cell is attractive because it is theoretically very efficient. Practical cells have achieved an efficiency of 47% but higher efficiencies are possible. The high temperature makes the cell insensitive to the presence of carbon monoxide that can poison the catalyst in lower temperature cells[2] and the operating temperature is high enough to carry out the reforming of natural gas internally within the cell. The high operating temperature means that there is high grade heat available for CHP. The commercial units that are available do not capture heat but there have been attempts at adding a bottoming cycle to the fuel cell to generate more power. The complexity of the MCFC means that it is not economical in small units. Typical unit size is from 300 kW upwards.

The solid oxide fuel cell. The SOFC is potentially the most attractive of all the fuel cells. Its electrolyte is a solid, an oxide material that will conduct oxygen ions through its structure when the temperature is elevated, usually to around 1000°C although some cells can run at lower temperatures. Electrodes can be attached to the solid electrolyte using microchip technology and the cells are extremely durable with lifetimes that, in principle at least, will exceed those of all the other technologies. The very high temperature means that natural gas reforming can take place within the cell. Practical SOFCs can achieve 50% efficiency and higher efficiencies should be achievable. The high temperature also makes a range of hybrid options including topping cycles and bottoming cycles feasible to boost power output. There have been attempts to commercialize both small units for distributed generation and utility scale power plants based on the SOFC but the greatest success has been achieved with small domestic CHP units which use the SOFC to provide electrical power. The main markets for these units have been Japan and Europe.

---

[2]Reforming of natural gas can lead to carbon monoxide formation and this can be a problem in low temperature cells.

FUEL CELL EMISSIONS

Fuel cells that burn pure hydrogen and oxygen from air produce only water and heat and represent probably the cleanest fuel-based generating system available. In practice however, most fuel cells are designed to burn natural gas from which they do generate carbon dioxide emissions. The reforming of natural gas to provide hydrogen for a fuel cell is a form of combustion but takes place at a relatively low temperature, using a catalyst, so nitrogen oxides are not formed in significant quantities. The process can lead to carbon monoxide formation, but the catalysts in low temperature fuel cells are sensitive to its presence and it has to be removed. High temperature fuel cells can exploit carbon monoxide as a fuel.

FUEL CELL CHP APPLICATIONS

The main application for fuel-cell-based CHP systems is to provide heating and hot water in commercial, municipal and domestic settings. The units are clean, quiet, and relatively low maintenance, making them ideal for this type of application. The larger commercial systems are primarily based on PAFCs while alkaline fuel cells, PEMFCs, and SOFCs are all being developed for the domestic CHP market.

The two high temperature fuel cells, the MCFC and the SOFC are both capable of providing high grade heat for industrial processes or to generate high pressure steam. However neither has been developed to provide CHP although the potential exists.

Ultimately the future of the fuel cell is probably tied to the development—or otherwise—of a hydrogen economy in which hydrogen becomes a fuel replacement for natural gas. The fuel cell is extremely efficient when supplied with pure hydrogen rather than natural gas. If the availability of hydrogen were to become widespread, then the fuel cell would form one of the most attractive means of converting this fuel into electrical power.

CHAPTER 8

Nuclear Combined Heat and Power

Nuclear power is one of the most controversial forms of electricity generation because of its association with the technology that is used in nuclear weapons and because of the extensive environmental damage that can be caused by a nuclear accident. Safety concerns have resulted in a massive slowdown in nuclear generation since its heydays of the 1960s and 1970s. Today, while some nations such as China and the United Kingdom are still pursuing power programs, others such as Germany have decided to close all their nuclear reactors.

Technically, a nuclear power plant is a thermal power station that relies on a heat engine, the steam turbine, to convert heat energy into electrical energy. The difference between nuclear power and more conventional thermal power plants is that the heat energy is generated through the use of a controlled nuclear reaction. In commercial nuclear power plants the controlled nuclear reaction is a fusion reaction in which a large atomic nucleus is split into two or more smaller nuclei with the release of large amounts of energy. Nuclear fission, the reaction that provides the energy in the sun, can potentially be exploited too, but the technology has not yet been developed to a stage at which it can be used commercially.

The heat from the nuclear reactor is used to raise steam to drive a steam turbine. The temperature of the steam generated by the main types of nuclear power plant is between 275 and 325°C, lower than for a coal or gas fired power plant and overall efficiency is lower too. The technology can be adapted for combined head and power (CHP) applications and heat from the reactor can be used to provide hot water for district heating systems. Residual heat from cooling towers of nuclear plants can also be used for horticulture. Nuclear CHP plants were used extensively in the Soviet Union and many of the countries that were part of the union still have nuclear CHP plants. The technology has been less common in other parts of the world but it has been used

Combined Heat and Power. DOI: https://doi.org/10.1016/B978-0-12-812908-1.00008-0

in Switzerland and Sweden. All the existing nuclear CHP plants are at least 30 years old. There have been more recent proposals for nuclear CHP projects but none has so far been constructed.

In addition to traditional nuclear technology, there have been a number of schemes put forward in recent years for the development of small, high temperature reactors that could be used to provide combined heat and power (CHP) for both district heating and for industrial processes. No plant of this type has yet been built for commercial use.

NUCLEAR TECHNOLOGY

Nuclear fission power plants generate their energy though the splitting of atoms with the release of large quantities of energy. The most important nuclear fission reaction that is exploited in nuclear reactors is that of uranium-235, one of the isotopes of natural uranium. Uranium-235 makes up only 0.7% of natural uranium but is the key atomic species for the generation of nuclear power.

Uranium-235 is naturally radioactive with a half-life of 704 million years. The isotope is also fissile which means it can sustain a nuclear chain reaction. When an atom of uranium-235 is struck by a neutron, its splits into two smaller atoms with the release of three further neutrons and a large amount of energy. This energy is carried away as kinetic energy by the three neutrons—the neutrons move at very high speed. Since each single atom that reacts in this way produces three new neutrons, these could each potentially cause three more uranium-235 atoms to split. This would lead to a cascade of fission reactions—a chain reaction. In practice, this will only happen if the piece of uranium is larger than a specific size called the critical mass. Otherwise, most of the neutrons will escape instead of stimulating further reactions. An atomic reactor contains more than the critical mass of uranium-235 but seeks to control this chain reaction so that it cannot run away. This is achieved by using various mechanisms including rods made of a material that can absorb neutrons. These rods are inserted into the core to reduce the rate of reaction and withdrawn to increase it.

When the nuclear reactor is active, the fission reactions release heat which must be carried away or it would cause the reactor core to melt.

To achieve this, a coolant is circulated within the core. This can be a gas or a liquid, depending upon the reactor design. In most reactors in operation around the world this coolant is water. There are also a small number that use carbon dioxide gas.

Two reactor designs predominate, globally. In the first, called a boiling water reactor, the coolant water within the reactor core is allowed to boil, generating steam which is used directly to drive a steam generator. The second, called a pressurized water reactor, operates with the core under high enough pressure that the water cannot boil. The hot, high pressure water is circulated from the core through a heat exchanger (often called a steam generator) where the heat is used to generate steam to drive the steam turbine.

The temperature of the coolant in the core of a water-cooled reactor is low compared to that in a coal-fired power plant. This results in relatively low pressure, low temperature steam being generated. Since efficiency of the heat engine, in this case a steam turbine, depends on the steam temperature and pressure, nuclear power plants are relatively inefficient, with overall conversion efficiencies of around 33%.

There have been attempts to design more efficient nuclear reactors that operate at much higher temperatures. The design that has attracted most attention is the high temperature gas cooled reactor that uses helium as its coolant. This can reach a core temperature of 900°C, leading to higher thermodynamic efficiency of the steam turbine. High temperature reactors of this type are much smaller than traditional reactors and are intended to by constructed as modular units that can be built in a factory. So far, however, no reactor of this type has been operated commercially.

NUCLEAR CHP CONFIGURATIONS

The exploitation of a nuclear power plant for CHP involves the extraction of heat from the plant, heat that can be used to provide hot water, normally for district heating. The heat is usually taken from the steam turbine although it could also be taken from the exit of the steam generator before the steam enters the steam turbine. In some cases residual heat from the cooling towers of the power plant has also been in horticulture.

Most nuclear power plants are extremely large, typically 1000 MW or more in generating capacity. These plants have large, compound, condensing steam turbines designed to extract the maximum energy from the relatively low pressure, low temperature steam provided by the reactor. The compound steam turbines will comprise a high pressure turbine and one or more intermediate pressure and low pressure units. Steam is normally taken from the low pressure turbines or between the high pressure and low pressure turbines. Wherever it is extracted, taking steam from the turbine will reduce the amount available for power generation.

In Russia, the country where nuclear district heating has been most widely used, the addition of the heating circuit has always taken place after the plant has been built, by adapting an existing plant. This has meant that an existing steam system has had to be modified for heat extraction. In these plants steam has been extracted from the low pressure steam turbines. Taking steam from the exit of the steam generator was examined but not used.

Switzerland has also used heat from three of its nuclear plants. However the amount of heat supplied has been limited. At the Muhleberg plant, the steam is extracted between the high pressure and low pressure turbines. Elsewhere residual heat from the cooling tower of a plant is used in a garden center.

There have been a limited number of proposals for modern plants of this type in other countries. One, in Finland, would also have extracted steam from the steam turbine. Discussions around the design of this plant included the economies of modifying an existing turbine design rather than designing a new turbine system.[1]

Aside from the technical issue of steam extraction, one of the key issues when considering nuclear CHP is safety. The nuclear reactions taking place within the core of the nuclear reactor involve the generation of some extremely dangerous radioactive isotopes as well as the production of high energy neutrons that can potentially generate new isotopes from materials such as steel if they escape the reactor core. Isolating the reactor and the cooling circuits associated with it are therefore vital.

[1]Nuclear District Heating Plans from Loviisa to Helsinki Metropolitan Area, Harri Tuomisto Fortum. Paper presented at the joint NEA/IAEA Expert Workshop on the Technical and Economic Assessment of Non-Electric Applications of Nuclear Energy, Paris, France, 2013.

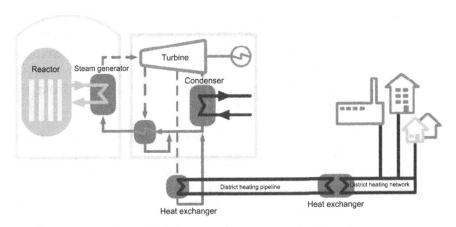

Figure 8.1 Nuclear CHP plant for district heating based on a boiling water reactor. Source: Fortum[2].

When taking heat from a nuclear plant steam turbine, there will normally be at least two additional heating circuits interposed. The steam from the steam turbine will pass through a heat exchanger where it will heat water in a secondary circuit. The heated water circulating in this secondary circuit will then pass through a further heat exchanger where it heats water in a tertiary circuit. The latter provides the hot water for the district heating system. A schematic of this arrangement for a boiling water reactor is shown in Fig. 8.1. To ensure safety in the event of a failure in any of these heating circuits, the pressure in the tertiary heating circuit is kept higher than that in the secondary circuit so that if one of the heat exchangers fails no water from the secondary circuit can pass into the tertiary circuit.

Safety considerations also dictate that no nuclear power plant should be built close to an urban area. In Russia, for example, the distance from a town or city varies between 25 and 100 km depending on the population density.[3] Similar considerations would apply in other countries. A consequence of this is that if a nuclear plant is to be used to supply district heating, the heating water must be transported over long distances. In Russia, for example, heat is piped from one nuclear plant to a town that is 40 km from the power station.

[2]Fortum presentation at NEA/IAEA workshop 2013.
[3]It appears that some plants have been built closer than 25 km too.

A proposed project that would have supplied nuclear heat to the Finnish capital Helsinki would have required a heating pipeline of around 75 km. This would have needed up to seven pumping stations as well as heat accumulators to manage the heat supply to the city. An alternative to the pipeline was to transport heated water from the power plant to the city by sea in vast containers.

Traditional nuclear power plants have limited ability to provide heat for industrial processes because the steam cycle temperature is generally too low for this to be effective, although in one instance in Switzerland heat has been used for paper processing. It would in principle be possible to provide process heat from one of the various high temperature modular reactors that have been proposed. The economic viability of this type of reactor has yet to be established.

NUCLEAR CHP APPLICATIONS

The use of nuclear power for CHP was first developed in Russia at the end of the 1960s as a means of reducing the demand for fossil fuel. The technology was later installed at plants in countries of the former Soviet Union. In all the instances this was a result of the adaptation of an existing nuclear power plant for power and heat. No nuclear installations were designed specifically for CHP. In 1989, the total heat capacity supplied in Russia from these nuclear plants was around 3000 MW_{th}.

The Russian nuclear industry supported two types of nuclear reactor, the RBMK gas cooled reactor and the VVER pressurized water reactor. For a typical 440 MW VVER reactor, the amount of heat extracted was 58 MW_{th} with heating water supplied at between 70 and 130°C.

The amount of heat extracted from each plant in Russia is relatively modest. In contrast, the proposed project in Finland cited above involved transport of up to 1000 MW_{th} from the nuclear plant with a nameplate capacity of between 1200 and 1700 MW of electrical generating capacity. The loss of electrical output from this plant as a result of heat extraction was expected to be around 17% or 200 to 280 MW depending on the design. The Finnish plant was not built.

During the 1980s the Russian nuclear industry expanded its interest in nuclear heating and began work on the design of a nuclear plant that was intended to provide heat alone. Designated the AST-500, each unit would have been able to provide 500 MW_{th} for district heating. Two pilot schemes were started, one in Gorky and the other in Voronezh. Both were abandoned after 1989 without being completed.

Interest in the development of nuclear CHP is limited outside Russia. There are four small projects in Switzerland and one in Sweden. There has been greater interest in the development of high temperature modular reactors that could provide both power and process heat. The USA has provided a site where high temperature reactor pilot modules can be tested, and there are several projects under development in the USA, Europe, and Asia. High temperature reactors often operate at high pressure and can drive a topping cycle gas turbine, while heat is extracted from the gas exiting the turbine exhaust. Many other configurations are possible too. However, it has so far proved difficult to make an economic case for any of the designs.

CHAPTER *9*

Renewable Energy Combined Heat and Power

The generation of electricity from renewable sources of energy such as the wind, the sun, or hydropower is not normally associated with combined heat and power (CHP). While these and other renewable resources are responsible for increasing amounts of electricity generation, their ability to provide heat directly is often limited. This presents a problem. According to most predictions, the long-term future of the global energy supply is based on renewable sustainable resources that do not damage the environment. In this future, electricity will probably be the key energy source. At the same time a large part of global energy is consumed in the form of heat, for cooking, for space heating, and for hot water. If the world is to switch to clean, energy sources that do not pollute the environment or increase the atmospheric burden of carbon dioxide, then some way must be found of generating heat from these sources too.

One simple solution to this conundrum is to use electricity directly as a heat source. In a world where there was an abundant supply of cheap electricity generated from renewable resources this would be an obvious solution and besides, electric heating is already used widely. Another is to develop a hydrogen economy in which hydrogen is manufactured—probably by electrolysis of water using renewable electrical power—and then the gas is burnt for heating as well as being used for transportation fuel. Both are viable solutions to the problem.

The alternative is to utilize renewable resources directly for heat and power production. Three renewable energy sources lend themselves to this, the sun, geothermal power and biomass. All are used today as heat sources. Traditional biomass provides one of the major energy sources for around 2.7 bn people around the world according to the International Energy Agency.[1] Solar energy is used in limited quantities for rooftop solar water heating and geothermal energy has been used for district heating schemes for many years in some parts of the world.

---

[1]World Energy Outlook 2016, International Energy Agency.

Combined Heat and Power. DOI: https://doi.org/10.1016/B978-0-12-812908-1.00009-2

In addition to these, solid waste material can be used in CHP schemes. Waste sources such as municipal solid waste may be considered renewable in some jurisdictions and waste to energy plants often combine the production of electricity with the supply of heat either to an industrial process or for district heating.

If the world is to progress toward an energy economy that is independent of fossil fuels, then it will be important to find the most efficient ways of using all these alternative resources. That will include the efficient exploitation of the three obvious heat sources as well as looking at other renewable resources in different ways.

BIOMASS CHP

Aside from fossil fuels, biomass is the most widely used fuel, globally. According to the World Energy Council's *Survey of Energy Resources*, it accounts for roughly 10% of world energy consumption. Most of this is in underdeveloped regions of the world where wood is the traditional resource used for cooking and heating. Biomass is also used to generate electric power. Generating plants burning biomass have traditionally relied on waste wood of similar products but many modern plants are being designed to burn wood that has been grown specifically for use as a combustion fuel. Biomass can also be provided as liquid fuel suitable for burning in a piston engine power plant. Both the traditional combustion plant and the piston plant can be used as CHP plants.

One of the key questions when considering biomass as an energy source is availability. There are many agricultural sources of biomass waste such as straw, sugar cane bagasse, rice husks and forestry waste which can be exploited in power plants. However, wider scale use of biomass requires energy crops such as grasses and fast growing woods that are raised specially as combustion fuels. These crops will often compete for land with food crops, and this can have serious environmental consequences, particularly if it leads to food scarcity or rising food prices. The same arguments apply when considering liquid biomass fuels. If biomass is obtained from a sustainable resource, and if it does not have a deleterious effect on food production then it can offer a valuable source of both electricity and heat. Where there is likely to be competition, competing demands need to be balanced carefully, and sensitively.

A traditional biomass power plant is a combustion plant that burns the biomass to generate heat which is used to raise steam and drive a steam turbine. Plants of this type are similar in concept to the steam turbine CHP systems discussed in Chapter 5. However, they tend to be much smaller than typical fossil fuel plants, often with generating capacities of 50 MW or less. This has an impact on their overall efficiency for electricity generation and many will operate at less than 30% efficiency. When used in a CHP configuration this low efficiency is not a major handicap. Modern biomass plants can be much larger (up to 300 MW) and provide higher efficiency but they also require large volumes of biomass fuel. Today, plants of this size are scarce.

In a biomass steam turbine CHP plant, steam can be extracted from the exhaust of the steam turbine (a back-pressure steam turbine) or from part way through the steam turbine to obtain steam at an intermediate temperature and pressure (extraction steam turbine). The actual configuration will depend upon the application with extraction able to offer steam at higher temperature and pressure, suitable for industrial processes such as paper manufacture. A back-pressure steam turbine may be more suitable if heat is required for heating and hot water.

Typical is the biomass district heating plant in St Paul, Minnesota, USA. This plant has a gross generating capacity of 33 MW and can provide up to 65 MW$_{th}$ of heat at the same time. The plant burns local wood waste that has been converted into chips, with power exported to the grid and supplied to local users, while heat is used to provide energy for a district heating system. The plant burns 250,000 tonnes of wood chips annually.

For smaller scale CHP, other approaches are possible. One is to use a gasifier to convert the wood into a combustible gas. This can then be burnt in gas-fired reciprocating engine to provide power with heat recovered from both the engine and from the gasification process used to generate hot water.

Liquid biomass fuels can be produced from a range of oil-seed crops such as rape seed or sunflower seeds. These fuels can be burned in piston engines that have been adapted for such fuels. These engines are usually similar to diesel engines and the fuel is often marketed as bio-diesel. The engine drives a generator, while heat is captured from the engine exhaust and engine cooling systems and used to heat water. Engine based systems of this type can be used to supply heat and power for small commercial, municipal, or residential customers.

Another source of energy for biomass CHP is landfill waste sites. When organic material that makes up much of municipal waste is buried in a landfill site it ferments and produces a methane-rich gas. This gas can be captured and used as fuel for a reciprocating gas engine. There are many plants of this type across the world and they often produce both electric power and heat for district heating systems. Individual engines can be up to 6 MW in electrical generating capacity and larger plants can be constructed using several engines.

With a range of technologies available, biomass CHP installations are becoming popular in the developed world as alternative sources of energy. However, their ultimate contribution to the global energy system is likely to be constrained by the quantity of biomass fuel that is available.

GEOTHERMAL CHP

Geothermal energy is energy that emanates from the core of the earth. This is an enormous resource but one that is usually difficult to exploit for electricity production. Those geothermal power plants that have been built exploit heat that is contained in underground reservoirs of hot brine that have been heated by energy from the earth's core. Good resources of this type are relatively limited. The largest number of power stations based on this resource is in the USA, and there are large installed capacities in Indonesia and the Philippines.

While underground hot reservoirs offer the easiest geothermal resource to exploit, others are available. The most important of these is the energy in hot rocks that are close to the earth's surface. This energy can be extracted by pumping water into the rock under pressure to fracture it so that water can penetrate within the rock structure and be heated by it. With a network of wells, hot water can then be pumped back to the surface from the hot rock. This technique has been tested but is not widely exploited. The other potent source is the magma found below volcanoes. The molten rock that constitutes magma reaches very high temperatures but means to exploit this have not yet been developed.

For power generation the best resource is a deep geothermal reservoir, up to 2 km below the surface. These reservoirs often contain relatively hot brine; the temperature can reach 350°C or more. If pumped to the surface, this brine is hot enough to generate steam when pumped through a heat exchanger and the steam will drive a steam turbine. The generation efficiency is relatively low but since the energy resource is free, the economics of these plants is favorable.

The geothermal resource can easily be exploited for CHP by cascading the extraction of energy. The highest grade energy is first used to provide electricity. Heat remaining in the geothermal brine is then used to heat water for space heating. For CHP to be effective, the brine must still be relatively hot when it exists the steam generator and this will limit the electricity system efficiency. However, overall efficiency will be high.

Geothermal power plants are common in Iceland and plants there will often supply both electric power and heat. For example, the Svartsengi plant generates around 47 MW of electrical power and provides 200 MW_{th} for a district heating system. The temperature of the brine which is extracted from the local geothermal reservoir is 240°C. Another Icelandic plant at a site where the reservoir brine temperature is 380°C provides 210 MW for the local grid and 290 MW_{th} for a local district heating system.

The Icelandic geothermal reservoirs are relatively hot and offer a good quality resource. The brine in many reservoirs does not reach such high temperatures. A town in Austria called Altheim has a district heating system that exploits water from a geothermal reservoir that reaches only 104°C. The initial geothermal well was used solely for district heating but when a second well was built, a small power generating system was introduced too. The low temperature of the geothermal brine makes it unsuitable for steam generation but it can be exploited by an organic Rankine cycle (ORC) turbine, a device identical in concept to a steam turbine that uses a low boiling point organic working fluid in place of water. At Altheim, the ORC generator has a generating capacity of 2.5 MW.

Geothermal CHP is attractive where there are easily exploited underground reservoirs. These are limited, geographically. In 2016, there was around 13 GW of geothermal generating capacity operating around the world according to the Geothermal Energy Association, and this exploited 7% of the known global geothermal resource. Even if all the known resource was exploited, it could only provide a small fraction of global heat and power demand.

SOLAR CHP

The sun is the earth's most important energy source. It provides the energy that fuels photosynthesis, providing all the food that is eaten by humans and animals as well as being the original source of all the fossil fuels that are used today for heat and power. The heat energy from the sun powers the earth's weather systems, bringing rain and driving the winds. It also heats the oceans and hot tropical oceans can be used to generate electricity in ocean thermal energy conversion plants. More importantly from a generation perspective, solar energy can be exploited directly either using solar photovoltaic devices, solar cells, or solar thermal power plants to generate electricity.

Where does CHP fit into all this? The answer today is that it doesn't. That does not mean that it could not in the future. The two types of solar electricity generating plants can both, potentially, be adapted for CHP. Take as an example a rooftop solar cell system. The solar cell exploits energy from the visible and ultra-violet part of the solar spectrum but usually leaves the infra-red, the heat energy untouched. In some cells, this energy simply passes through the cells which are transparent to it. In others it is absorbed, heating the devices which must then be cooled. At the same time solar heat is often used for rooftop solar water heating. In principle, it should be possible to integrate the two so that a rooftop panel could produce both electrical energy and hot water.

The other type of solar power station, the solar thermal power plant, uses the heat of the sun directly. These plants are similar to more conventional power plants, with a solar heat collection system providing the heat input for the steam generator in place of a combustion furnace or a nuclear reactor. Large plants of this type use the heat to generate steam and drive a steam turbine and this type of cycle can be adapted in exactly the same way as a conventional steam turbine plant to create a CHP plant. There may be logistical problems with this approach since many solar thermal power plants are sited in remote regions where there is high solar insolation but not many people. In principle, however, the two can be combined.

The main issue is cost effectiveness. It is relatively simple to exploit the sun directly for hot water and space heating. Good architectural design (and traditional design too) will exploit the sun to maintain ambient temperature in buildings. There is also the problem that heating is needed in colder regions of the world or during colder seasons where sunlight is scarce, not in those regions and seasons where there is ample sunlight. Rooftop CHP units may eventually prove economical but other solar CHP configurations may prove less attractive.

CHAPTER 10

The Environmental Implications of Combined Heat and Power

Combined heat and power (CHP) is an approach to energy production that can significantly increase the efficiency of energy use. When implemented, it can reduce the need for new power generating capacity and hence reduce the overall environmental impact associated with the production of energy. In most cases, the additional cost should be minimal because the technology utilizes energy that would otherwise be wasted.

As has already been outlined in earlier chapters, many of the technologies that are used to generate electricity produce large quantities of waste heat. With a few exceptions, generating plants that burn fossil fuels or other combustible fuels convert less than half of the energy from the fuel into electricity. The rest emerges from the process as heat. At the same time, there is a massive demand across the globe for space heating, for hot water and for industrial heat. If waste heat from power stations can be used to provide this heat, then the need for consumption of additional fuel is avoided. This also avoids the production of a wide range of pollutants and emissions including carbon dioxide from fossil fuel combustion.

For logistical reasons, not all power stations can be adapted for CHP use but many can. The International Energy Agency estimated in a report published in 2008 that widespread use of CHP could reduce global carbon dioxide emissions by 10% by 2030. This would also reduce the investment needed in the power sector infrastructure.

COMBINED HEAT AND POWER AND LOCAL HEAT SUPPLY

The technologies that underpin CHP are all well established. Exploiting them involves little additional risk. However there are significant barriers to widespread implementation. One of the major barriers relates to the expanded use of district heating.

Combined Heat and Power. DOI: https://doi.org/10.1016/B978-0-12-812908-1.00010-9

One of the biggest uses of heat is for domestic and office heating. In consequence, space heating presents perhaps the greatest opportunity for savings when implementing CHP. For the majority of homes and offices in the developed countries, space heating is supplied from an individual domestic or office heating system, often burning a fossil fuel such as natural gas. If this heat could be supplied instead from a communal district heating system using waste heat from a power station then significant energy savings could be made.

Not all homes and offices can be supplied in this way, but those in towns and cities are ideally suited to this type of heat delivery. Unfortunately, however, the installation of a district heat distribution system in a city where one does not already exist involves massive upheaval and expense. Most city authorities will not contemplate such a major infrastructure project except in situations where new housing is being built. Since most of the world's towns and cities do not have district heating systems the potential for expansion of this type of system appears limited. From an environmental perspective, this is a lost opportunity. There are other opportunities, however.

The exploitation of large central power stations for CHP is often limited by their location. Many are remote from high population density centers where their waste heat would be needed. Heat can be transmitted by pipeline over long distances, but this is not always considered either economical or practical. Distributed generation, where small generating units are located closer to the customers, is much better suited to CHP. A small reciprocating engine can provide both heat and power to a residential development, a hospital or a small commercial facility. Again, this requires the use of communal rather than individual heating but implementation at this level is much easier. This too has large environmental benefits in terms of lower atmospheric emissions and more efficient use of both heat and electricity.

COMBINED HEAT AND POWER AND RENEWABLE ENERGY

While CHP has obvious benefits in terms of energy use, the long-term future of global electricity supply will be closely linked to renewable sources of energy that are both sustainable and do not pollute the environment with their emissions. While some of these renewable technologies are capable of supplying both heat and power the most important, hydropower, wind power, and solar photovoltaic power are not sources of heat. Solar cell installations might be adapted to provide heat as well as electrical output but the heat production would be separate from electricity generation. Wind power and hydropower offer no opportunities for heat production.

There is one other source of electricity that does not emit carbon dioxide into the atmosphere, nuclear power. The nuclear industry is keen to promote this as a clean source of both electricity and heat but costs and environmental concerns are likely to limit its appeal.

If renewable generation does come to dominate the energy industries, then utility of CHP will be massively reduced. When electricity generation does not rely on heat engines converting heat from combustion fuels into electricity, then the amount of waste heat available is limited. There will still be some applications in industry where combustion processes are used but most opportunities for CHP will disappear.

This will take a long time. Fossil fuel will still be providing as much as half of the world's electric power in the middle of this century and its use is likely to continue at least to the end of the 21st century. During that period, several generations of CHP plants can come and go. But eventually, the technology will become obsolete.

COMBINED HEAT AND POWER AND EMISSIONS

A CHP plant utilizes the waste heat from a generating system that uses a heat engine to convert heat energy into electric power. (Fuel cells are the exception to this generalization.) In each case, there is an underlying technology that is used to generate electrical power from a fuel and this remains unchanged when heat recovery is added. The environmental impact of CHP will therefore depend on the environmental impact of that underlying technology.

Carbon dioxide emissions are one of the main concerns today. Carbon dioxide is produced when fossil fuels and biomass fuels are burnt to generate heat. Production is inevitable during the combustion of any carbon-containing fuel in air. However the relative amount produced depends on the efficiency of the process being exploited.

The normal way of assessing the relative level of greenhouse gas emissions from a power plant is to compare the amount of carbon dioxide released for each unit of electricity generated. Using this metric, it is clear that if two power stations burn the same fuel, then the one that converts more of that fuel into electricity—the more efficient—will produce more electricity for each unit of fuel it consumes and therefore less carbon dioxide per unit of output.

CHP increases the overall efficiency of a power plant or power generating unit because it uses waste heat energy as well as generating electricity. However, the increase in efficiency cannot be measured in terms of a decrease in the amount of carbon dioxide released for each unit of electricity. In fact, the amount will probably increase because CHP plants are often less efficient at generating electricity than dedicated power plants.

While it is true to say that the amount of carbon dioxide released for each unit of energy used (when the energy exploited for heating is added to the electrical energy produced) is smaller, a better way of assessing the impact of CHP is to calculate the avoided carbon dioxide production resulting from implementation of CHP. In other words, how much fossil fuel would have been burned to provide the same heat as that provided by the CHP plant. This is the type of calculation that underpins, for example, the IEA calculation cited above that carbon dioxide emissions could be reduced by 10% by 2030.

The impact of CHP on carbon dioxide emissions measured in this way is significant. In the same way, CHP will reduce the emissions of other pollutants that result from power generation. So, for example, if a coal-fired power station is converted to CHP, then the amount of sulfur dioxide, nitrogen oxides, and particulate emissions will be reduced if the heat from the plant replaces the combustion of coal elsewhere.

Emissions will be reduced in this way, but not eliminated. A coal plant that provides heat and power will emit more or less the same quantities of these various pollutants as a similar plant without CHP. The same goes for gas-fired plants and all the other types of plant that rely on a fossil fuel or combustion fuel. Even biomass plants produce atmospheric emissions and must be evaluated in the same way. CHP is therefore a way of reducing the environmental impact of power plants, not of eliminating it.

CHAPTER 11

The Economics of Combined Heat and Power

Assessing the economic viability of a combined heat and power (CHP) plant involves estimating the capital cost of installing the facility, the cost of the fuel it will use, and the cost of operating and maintaining it over its lifetime, then comparing this to the equivalent costs associated with a stand-alone power generation unit (or the cost of buying the power from the grid) and the cost of providing heat from a local energy source such as natural gas or oil. In some cases, there may be tax incentives associated with the adoption of CHP to be taken into account and some jurisdictions offer grants against the installation cost too.

In principle, these factors appear simple to evaluate but as the earlier chapters in this book have shown, there are a wide range of CHP plant configurations to choose from. Each of these has its advantages and when CHP is an option the choice between them may not be easy. Any decision is likely to depend on a variety of factors that will include the capital outlay for the installation, convenience and the estimated savings. Personal preference and prejudice may also come into play. Persuading a group of occupants of a residential block to adopt communal heating instead of individual heating systems may prove difficult even if the economic advantages seem clear. Independence may be valued above economy.

With so many different systems to choose from, a systematic analysis of costs is beyond the scope of this volume. Instead, some cost figures are presented below which can be used to provide broad guidance on the most significant cost elements.

CAPITAL COSTS

Table 11.1 contains some representative capital costs for CHP systems at the end of the first decade of the 21st century. Some are in pounds sterling and some in US dollars.[1] As the table demonstrates capital costs vary widely.

[1]A reasonable but rough conversion can be carried out putting one pound Sterling equivalent to one and a half US dollars.

Combined Heat and Power. DOI: https://doi.org/10.1016/B978-0-12-812908-1.00011-0

Table 11.1 Some Representative CHP System Costs

System Type	Estimated Lifetime (years)	Cost
Small natural gas fired reciprocating engine (<15 kW)	5–6	£3290/kW ($5000/kW)
Large natural gas fired reciprocating engine (110 kW electrical output)	10–20	£890/kW ($1300/kW)
Gas turbine CHP (100 kW electrical output)	10–15	£900/kW ($1400/kW)
Steam turbine CHP	20	>$2000/kW
Phosphoric acid fuel cell (400 kW electrical output)	10–20	$2500/kW
Microturbines	10–20	$3000–4000/kW
Biomass CHP	28	$5800/kW

Source: Greenspec[2], Datamonitor[3], NREL[4].

When fossil fuels are the main energy source, a small reciprocating engine-based system of less than 15 kW is the most expensive configuration in the table with a cost of around £3290/kW or close to $5000/kW. Small CHP gas engines tend to be expensive but the cost falls as the size increases so that a similar system but with an electrical generating capacity of 110 kW instead of less than 15 kW has an estimated cost of £890/kW ($1300/kW). This is comparable to the cost of a similarly sized gas turbine-based CHP system which, according to the data in Table 11.1, would cost £900/kW ($1400/kW). These two large CHP systems would be expected to have a lifetime of up to 20 years but the small reciprocating engine operating at around 5000 h each year or more would have a lifetime of only 5–6 years.

Other systems included in Table 11.1 are a steam turbine-based system with a cost of in excess of $2000/kW. This price is for a large steam turbine-based plant and the cost will rise for smaller systems. A commercial fuel cell based on phosphoric acid fuel cell technology is available in the USA for around $2500/kW. Meanwhile, microturbines are still relatively expensive at $3000 to $4000/kW.

The table also contains an estimate for a biomass CHP plant. The cost quoted in the table is $5800/kW. This is higher than that for any

[2]CHP, *Peter Mayer of Building LifePlans, Greenspec.*
[3]The Future of Distributed Generation *and* The Future of Fuel Cells, *Paul Breeze, Datamonitor, 2009 and 2012.*
[4]*Distributed Generation Renewable Energy Estimate of Costs, (updated February 2016), US National Renewable Energy Laboratory.*

of the fossil-fuel plants and is a result of the lower efficiency of biomass plants. However the cost does depend on size and large biomass plants may be cheaper than this. A separate cost estimate from IRENA found that the cost of a biomass CHP plant with a stoker boiler was $3550–$6820/kW while similar costs for a gasifier CHP plant were $5570–6545/kW.[5]

To obtain the best economic return it is important to size a CHP system correctly. The unit needs to operate for at least 4000 h each year to be cost effective, a particular concern where smaller systems are under consideration. Oversized engines running at less than full output will often lose some economic benefit, particularly if this involves dumping heat because it is not required.

FUEL COST

The cost of fuel plays an important part in the overall cost of energy from a CHP plant. Most modern urban CHP plants burn natural gas which is a globally traded fuel and the price varies through cycles. The cost is currently low, particularly in the USA where the onset of shale gas production during the last decade has brought prices down considerably. Natural gas prices can display extreme volatility and this adds a risk element to the selection of natural gas as the fuel for a CHP plant. Coal is normally cheaper than natural gas but will rarely be used in a modern CHP plant. There are examples of older large municipal plants that burn coal as well as some industrial plants. There may be a significant installed capacity of these types of plant in the main coal bearing regions of the world, such as China of in some eastern European nations.

Biomass is a modern fuel choice. Depending on the source of the fuel, biomass can be relatively cheap. The cheapest source is waste material, particularly wood waste. For modern usage this is normally formed into chips. Some municipal waste can be sorted and turned into a fuel suitable for CHP combustion too. Wood chips that come from specially grown crops or from forests will be more expensive. However the prices are likely to be relatively stable, particularly if long-term contracts can be struck.

[5]Biomass for Power Generation, International Renewable Energy Agency, 2012.

COST OF ELECTRICITY

The cost of electricity from a CHP plant will be one of the most important factors in determining the economics of the plant. This cost will depend on both the capital cost of building the plant and the cost of fuel. Economics will also depend upon type of consumer that the unit is supplying. Large industrial consumers can buy electricity at wholesale prices from suppliers. The cost of the same electricity when supplied to a domestic consumer can be several times that of the wholesale price. Therefore, a domestic CHP system can be relatively expensive and yet still economically viable.

IRENA has published cost of electricity figures based on the economists' "levelized cost of electricity" (LCOE) model that is commonly used to make cost comparisons. It found that the LCOE from a biomass stoker boiler CHP plant was between $70 and $290/MW h while for a gasifier-based CHP biomass plant the cost of electricity was $11–$280/MW h.[6] Meanwhile, another study estimated LCOE from a 7.3 MW gas turbine CHP plant in the USA to be $62/MW h, while for a 21.1 MW plant, this fell to $54/MW h.[7] The International Energy Agency in its 2015 version of its *Projected Costs of Generating Electricity* report found the cost of electricity from CHP plants varied between $25/MW h and $70/MW h depending on economic and national factors.

While the cost of electricity forms an important element in the economic equation, the value of the heat that the plant provides must be taken into account too. It is normally the additional savings that accrue from this that make the CHP plant a cost effective choice. Evaluating this will be a matter of estimating the cost of the same heat from conventional sources. This will then be accounted a saving since the CHP plant negates the need for this outlay.

[6]Biomass for Power Generation, International Renewable Energy Agency, 2012.
[7]How Electric Utilities Can Find Value in CHP, Anna Chittum, American Council for an Energy-Efficient Economy, 2013.

Note: Page numbers followed by "*f*" and "*t*" refer to figures and tables, respectively.

Printed in the United States
By Bookmasters